'ISO 9001 certification just got easier. All you need to do is put structure to logic, and PMI shows you how.'
Adam Gade, Managing Partner at A&D Resources, Denmark; Former Vice President and Chief Information Officer, Maersk Line

Praise for Making Your Work Work

'*Making Your Work Work* reveals great knowledge of as well as love and humbleness towards the topic, and offers priceless value to the reader. Am I surprised? Well thinking of the numerous books I have digested during the last decade, I actually thought that all there is to write had been written, but no, this book works! I therefore have no doubt the huge impact *Implementing ISO 9001: 2015* will have – for those who follow the methodology and thinking – in achieving impressive results not only in terms of business improvement, but also in the pursuit and achievement of ISO 9001 registration in the future.'
Heidi Kaas, CEO at Quatre Thinking Process Aps, Denmark

'*Making Your Work Work* shows you how to manage your processes while simultaneously improving them. It doesn't offer a magic bullet and doesn't pretend to. What it does is demonstrate a tested, proven and scientific approach. Founded on Deming's Plan Do Study Act cycle (PDSA), the book takes you through all the stages starting with 'study', through to what to do when the improvement is recognised and you are challenged to lead to the next level. I've practised the basic tools and philosophy in this book for years in different industries, and I know that they work. The style is light but there is a lot of meat, and you'll be referring back to it often.'
Senior automotive industry engineer

'At long last a business improvement book that goes back to the basics taught to us by Deming. Yes there are Lean, Six Sigma, TQM, etc., but they all come from one essential source and that is the work of Deming. It's great that in this book there is a real insight into how a manager can think through problems and implement genuine improvements by looking at organisations and departments through just four simple concepts.'
Business improvement manager, UK Government Department

'Basing itself on fundamental principles this book gives excellent insight into how to think and behave as a manager and leader who wants to focus on improving their workplace and services to customers. Written in an accessible style it combines the 'story' of a manager learning and growing, alongside case studies and the application of theory into practice. It is certainly helping me develop my style and approach, and becoming a well-thumbed reference book as I help my organisation develop its thinking to see itself as a system and to continually improve and transform.'

Howard Davies, Head of Continuous Improvement, Natural Resources Wales

This is for any business leader who wants to go beyond improvements and understand the underlying mechanics and systems that secures long-term effects. Having worked hard on the methods in the book *Making Your Work Work*, I keep it close as a reference guide. The fact of the matter is that the Deming system map is probably the single biggest epiphany I have had in business improvements and I keep coming back to that as a way to understand organisations.

Mads Bronden Wijngaard, Head of Business Transformation, Maersk Line, Denmark

Implementing ISO 9001:2015

Thrill your customers and
transform your cost base with
the new gold standard for
business management

Jan Gillett
Paul Simpson
Susannah Clarke

infiniteideas

First published in 2015 by
Infinite Ideas Limited
36 St Giles
Oxford
OX1 3LD
United Kingdom
www.infideas.com

Implementing ISO 9001:2015 is an updated edition of Making Your Work Work, first published by Infinite Ideas in 2014.

A CIP catalogue record for this book is available from the British Library
ISBN 978–1–908984–50–0

Brand and product names are trademarks or registered trademarks of their respective owners.

Typeset by KerryPress
Printed in Spain

Contents

Foreword

The revision to ISO 9001 is due to be published in September 2015. I am very pleased that my friends Jan Gillett and Paul Simpson and Jan's colleague Susannah Clarke have co-authored a guidance book for the revision.

I am a believer in the Japanese way of Total Quality Management (TQM) and have been researching and teaching the theory of quality and its application to management at the University of Tokyo for many years. I have also been involved in development of the ISO 9000 series of standards since the mid-1980s and Jan and I have discussed quality enthusiastically for many years in a quality group, GQFW (Global Quality Future Workshop). Paul is also a colleague from our work in a task group revising the Quality Management Principles (QMP) in ISO/TC176, for which I was the leader and he was a core member.

The reason why I am so pleased that they have authored the book is not only because they are friends and colleagues, but also because the 2015 revision is intended to promote a purpose-oriented way for design of a Quality Management System (QMS). ISO 9001 has always been used for QMS certification. If we recognise that quality is the customer's evaluation of the value provided through products and services, the management system model should be a critical model for business management.

The most important revision is, in fact, in clause 4 (context of the organisation), which requires organisations to design and establish their QMS autonomously. If they respond to changes clause by clause, the assessments in the certification process will focus on trivial details and come apart from their business. The problem could be a lack of consideration on objectives of QMS.

Regrettably in QMS certification, superficial conformance to ISO 9001 requirements is the norm. The three authors have a different view: they advocate that organisations should understand their QMS as a system and make the best use of it for business management. So, when they talk about 'system', they firstly think of the system as consisting of many elements and secondly that the whole system has objectives. By determining system objectives, identifying their elements, and understanding relations between these objectives and elements the system can be optimised.

The model for QMS in ISO 9001:2015 exactly follows this thought. The revised ISO 9001 is a model to design, establish, implement and improve the QMS as a system with the objective of customer satisfaction and consisting of processes. We should determine the QMS objectives, the capability required to achieve these objectives, install the capability in QMS elements, and continue to improve the QMS.

This book explains how to make effective use of the ISO 9001:2015 revisions, based on total understanding of the profound meanings incorporated in the revised QMS model. I do strongly recommend stray sheep in the field of QMS certification to experience and enjoy this book. You will be awakened.

Yoshinori Iizuka, Professer Emeritus, The University of Tokyo,
August 2015, in a GQFW meeting held in Winnipeg, Canada

Introduction

'*Thrill your customers and transform your cost base with the new gold standard for business improvement.*'

Or how about:

'*Integrate transition to the new standard with improvement and transformation, and get your investment back many times.*'

These are strong statements to make, but we are confident that you will agree they are valid when you have applied the approach we describe.

The purpose of our book is to help readers understand the key aspects of the revised ISO 9001:2015 standard, and to appreciate how this approach can help in improving as well as assessing or auditing their organisation to the new Standard. In Chapters 1 to 6 we have included the clause numbering and structure from ISO 9001 with a summary of what the standard requires organisations to do to implement them. The words of the standards are not used for two reasons: standard text is copyright to ISO and national standards bodies, and standard wording is occasionally archaic and terms are selected to be as unambiguous as possible in all languages used by the 160 participating organisations. The wording used in this book should be more accessible for readers of this book as a consequence.

Both this book and ISO 9001:2015 have to be linear representations of an organisation's quality management system and so cannot easily capture the complexity and interrelationship of an organisation and its context. Our chapters and clauses in ISO 9001 are logical groupings of activities and requirements but you will often need to consider activities from multiple chapters of the book and a mix of ISO 9001 clauses simultaneously. Nonetheless, the sequence we have taken leads naturally to achieving the organisation's goals and producing a robust quality system.

Over a million organisations worldwide will have to make changes in making the transition to the 2015 version of ISO 9001, and some may anticipate that the expense is not going to generate much of a return on the investment. However the changes can be used to stimulate transformation in your organisation and its wider environment, coordinating its processes to achieve both its customer goals and reduced operating costs.

So how do you do that?

To get your return on your registration investment, you need to learn what is involved with improving how the work works in delivering tangible benefits **in parallel with** achieving registration. The challenge is not simple but the approach works really well when applied as a continuing transformation: diligently over an extended time.

Context for the changes

The 'quality improvement' movement can be traced back centuries. Its successful manifestations are in companies that have made it their way of life, starting in the 1950s with leading Japanese organisations, but since adopted by others in the rest of the world. When used properly the approach has achieved astounding and continuing transformations in cost reduction and customer delight. Leaders in these organisations will take the revisions to ISO 9001 in their stride. Unfortunately they are only a small proportion of organisations registered to the existing standards.

For all too many of the others ISO 9001 registration has been something that has to be attained but is seen as a distraction from the real work. For such organisations the revisions will create some big challenges, for themselves, and internal and external auditors. Their first challenge is likely to be the recognition that previous practice is not going to work in future.

Three particularly important changes are:

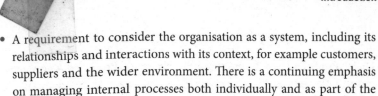

- A requirement to consider the organisation as a system, including its relationships and interactions with its context, for example customers, suppliers and the wider environment. There is a continuing emphasis on managing internal processes both individually and as part of the organisation's system.
- Requirements to proactively consider risk at a system, process, as well as product level, and to plan for changes to take advantage of the opportunities available, not just to mitigate the problems as they may present themselves.
- Requirements to plan for continuing improvement and transformation, not just problem solving.

The effect of the changes is to integrate third-party assessment with change, improvement and transformation. No longer will it be acceptable to audit paperwork and data; judgements on leadership performance are called for, together with diagnosis of what needs to change.

Process Management is thus coming of age

When PMI was founded in 1985 most people had no idea about managing work as a process: the idea was associated with chemical plants or computer programs. It was an uphill task even talking about processes with managers. But over the decades the process management concept has become increasingly widely appreciated, and indeed was incorporated in the 2000 revision of ISO 9001. However, it was still possible to go through the motions in getting registration. Even now, many organisations have not yet made the shift to process orientation, much less to relentless customer-focused transformation of their whole system as a network of interdependent processes.

The time has thus come for organisations to ensure that tangible strategic and operating benefits are delivered in tandem with the formalities of developing the Management System for ISO re-registration.

Dr Deming's System of Profound Knowledge provides the principles

Dr W. Edwards Deming proposed his System of Profound Knowledge,[1] towards the end of his amazingly extended practice in organisational transformation. Building upon his experiences throughout the world from the late 1920s to the early 1990s, he identified four interconnected fields that require a level of working knowledge by every manager, as demonstrated in this diagram. It is our experience that the model provides the most robust set of foundations of any in the leadership field, being valuable in every application we have ever encountered.

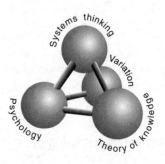

We have thus based our practice on the System of Profound Knowledge for over 25 years. We have worked in and studied organisations around the world, many of which have been awarded prizes including the Deming Prize, the Baldrige Award and the European Quality Award. Building upon that exposure, our vision expresses the universal essentials that organisations have to achieve to do well against these criteria, and is completely applicable to the assessment required by ISO 9001:2015.

The organisation is achieving its goals, and can demonstrate its management and improvement approach:

- **Everywhere.** Across the whole organisation, including strategy development, everyday work, projects and programmes.
- **Everyday.** Leaders understand and can explain the relationship between how they approach their daily work and the overall improved results they are achieving.

---

1   Described by him in *The New Economics* in 1993, and expanded in many of our recommended books

- By **Everyone.** The approach is used in depth where appropriate, and can be explained by line managers and staff routinely, not just the improvement personnel.
- **For Ever.** It has clearly been applied and developed over many years.

If your organisation is doing all these you can approach the new standard with confidence. The expectation of a registered Quality Management System is that, properly implemented, it will meet this vision.

A need for continual learning

Leading an organisation towards this state demands continual learning, by everyone, applied every day. The basis for learning taken in the ISO Management Systems Standards (MSS) including ISO 9001 is the PDCA (Plan-Do-Check-Act) cycle, often known as the Deming Cycle. This forms a major part of the Theory of Knowledge and has been fundamental to our approach for over three decades. We use PDSA (Plan-Do-Study-Act), as proposed by Dr Deming towards the end of his career. As with terms used in ISO 9001 the abbreviation and its meaning differ depending on your nationality and culture, as you can see from this example from Germany.

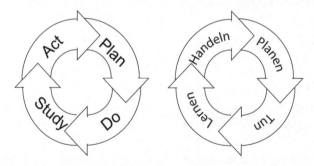

Even if you use a single language such as English, used in this book, terms can be controversial and require explanation to ensure the term conveys its intended meaning. The most obvious example of this problem is the word 'Check', used in the first version of the cycle developed by Deming, and some of the pioneering Japanese in this field, in the 1950s. This sector of the cycle concerns observation, exploration, testing, researching, and an awareness of degrees of accuracy, repeatability, bias and so on, whether the subject is an experiment, today's output or this year's strategic planning. In Western English the term 'Check' has connotations of box ticking and this led to its replacement with 'Study' to imply a more thoughtful assessment of current state. But 'Study' has its own detractors with images of students cramming to prepare for exams. When we, the authors, think of 'Check' or 'Study' it is with an organisation in mind understanding its current situation in context, clarifying its purpose, making plans for improvement, carrying out tests, watching how things are doing, and evaluating the results.

For these reasons you may find it helps to talk of the Deming Cycle, and stimulate a discussion on what each sector really means in your own language. We explore this further in Appendix 1.

Where are you now?

Your first step is to consider what you think about your organisation's current capability and performance. This will lead you to further questions depending on your circumstances and knowledge. We expand on this question in Chapter 1, but meanwhile you might like to consider these questions, depending upon your role.

The CEO

- To what extent does your quality system contribute to your organisation's purpose, vision and goals?
- How are your strategic goals realised through the application of your quality management principles?
- How well does your system provide you with the data you need about the strengths, weaknesses, opportunities and threats to your organisation so that you can lead and innovate effectively?
- As a Board, what is your collective opinion of your quality management system? Is it seen as an enabler or a necessary evil?
- How well is your organisation integrated with your key suppliers? Do they take ISO 9001 seriously? Can you learn from each other?
- How does your quality system enhance your organisation's ability to delight your customers?

- If you wanted to make a major change, acquisition or merger, how likely are you to seek the input of your quality team to help guide this process?

Top managers

- How well is your organisation achieving its strategic goals and what are the prospects?
- How do you know if all the important work (everyday processes and projects) in your organisation is capable of satisfying and delighting customers?
- How do you know if it actually does it? (satisfy and delight customers).
- Are you comfortable that all these activities are operated in standardised fashion, as far as is needed, by people who are competent and sufficiently qualified to carry it out?
- How is your quality system contributing to your organisation's everyday work and strategic development?
- What is your experience of external MSS auditors talking with you and your colleagues in diagnosing problems and facilitating solutions?
- Do you expect your internal audit function to constructively challenge top managers or leaders when they uncover issues and problems? How successful have they been in doing this?
- In relation to management and improvement, how consistent are people across your whole organisation and its supply chain in their philosophy, language and methods?

Auditors

- How well can you work with managers and people in everyday jobs to jointly understand the real state of processes and their alignment with documentation in the MSS?
- How do you interpret variation and use it to tell the difference between problems and abnormalities?
- How competent are you in 'Lean', 'Six Sigma' and other improvement methodologies so that you can tell if they are achieving their potential?
- Do you understand how to assess measurement systems to distinguish between variation in output and variation in measurement?

- How successful have you been in working with suppliers in a constructive way to understand their capability?

Improvement and change managers

- How satisfied are you with the level of support of top managers, and the resources they make available to do justice to the goals they have in mind?
- How would you rate your own skills to be a confident facilitator of all levels of projects and subsequent standardised implementations?
- Are you completely familiar with the details of ISO 9001:2015 to ensure that your contributions pave the way for a trouble-free audit whenever it may happen?

Managers of everyday work

- Are you confident in working with teams to understand what is going on, using your organisation's standard methods to improve it, and implement changes that would be consistent with your (quality) management system?
- How do you rate the relevance and disciplines of your organisation's management system to the actual job you have to do?
- How confident are you that internal assessors/auditors will be ready, willing and able to help when you need it?

Implications for your organisation's registration process

For leaders

This quote from IRCA[2] indicates the intentions of the authors of the standard, and of a new approach to audit.

> *The switch from Management commitment to Leadership and commitment. The changes around Leadership represent a deliberate attempt by the standard writers to embed responsibility for operation*

---

2   International Register of Certificated Auditors, www.irca.org

and performance of the QMS at all levels in the organisation. Previously it was not uncommon for the QMS to be focused on the management representative; however, the requirement for this role has now been removed. The Leadership changes have considerable implications for both top management and those who will be required to audit top management.

In some cases this may be no change from the current situation – if top managers see their role as being responsible for leading the organisation as a system, optimising its performance for customer benefit, and for the quality management system to be integrated in this purpose. Such organisations, which probably have an active self-audit programme, will find useful guidelines that support their existing practice.

But if the leadership has not been fully engaged, the implications are that quality professionals, internal auditors and external auditors will need to work much more with the top leadership team, the MD and/ or CEO. No longer is there a glass ceiling where the buck stops with the 'management representative'. There *is* a requirement for top management to demonstrate leadership and commitment to the quality management system through:

- their accountability for the effectiveness of the system;
- ensuring the policy is communicated, understood and applied within their organisation;
- promoting awareness of the process approach; and
- ensuring it achieves its intended results and engaging with the people to contribute towards the effectiveness of the system.

The section in Leadership, 5.1.2, Customer focus (summarised in Chapter 2) makes putting the customer at the heart of the organisation key to success with the Standard. Leaders have to be able to demonstrate leadership and commitment to customer focus, customer requirements, risk and opportunity, **consistently** providing products and services that meet customer requirements and enhancing customer satisfaction.

This section, if it is embraced as intended, offers a transformational opportunity for organisations to lead and commit to a focus on their customers, engage with and to engage their people in the quality of their products and services and to see the great results that they can deliver as a result of this.

Quality managers

The quality manager will be having conversations with their leadership team to increase readiness to change, in particular to capture their thoughts and expectations for strategic planning to cover organisational context (addressed in ISO 9001 Clauses 4.1, 4.2 and 4.3). They will need to explain the changes in emphasis for the Standard and what they as a leadership team will need to do in order to successfully comply. But of course it's more than just compliance. If the organisation is going to successfully reap the benefits of the new Standard then the whole organisation will need to engage in working in this new way: a process approach where everyone is contributing towards the effectiveness of the system, for the customer. From here, compliance to the Standard is inevitable.

Auditors

IRCA has this to say about new requirements of auditors:

> In order to be able to audit Context effectively, QMS auditors will need to have formulated a picture of the context of an organisation prior to carrying out the assessment. The auditor will not be able to conclude if the requirements surrounding internal and external issues, and the relevant interests of relevant interested parties, are being properly addressed unless they have an appreciation of what these should reasonably be. This will require additional planning using unfamiliar reference sources.
>
> Auditors must also understand the link between Clause 4 Context, Clause 6 Planning and Clause 8 Operation, and be comfortable auditing these interdependencies.

Auditors will need to adapt their approach to conducting an audit, how they observe the processes operating and what questions they ask of the organisation, its leaders and people. They will need to test top management's understanding and vision of their QMS through interview, before going to *gemba* (where the work is performed) and observing the processes in reality. This could be tricky for all parties. Organisations who may be familiar with the traditional audit approach could be resistant to this change and may object to being challenged by auditors about how they demonstrate leadership and commitment.

There are several aspects of the documentation that need to be reconciled to an organisation's approach. The more synergy that is found, the easier the process of registration should be. Further, whether the synergies extend across leadership attitudes and behaviours to include the style and the methodologies of everyday and strategic management.

Clearly a lot depends on how the auditors currently audit, as with all roles there is a degree of variation. The feedback we have had from the auditors we have talked to is that there are potentially two primary areas that the majority of auditors will need to develop:

- **Skills competency** – especially in the soft skills such as influencing, engaging others, dealing with conflict, negotiation, and change management from the social emotional perspective.

  External auditors, who will find themselves meeting with senior leaders and questioning them about their system, its performance and the context of their organisation, may well need to hone their skills in preparation for the reactions or responses they may get.

  Similarly it's not unusual to hear an internal auditor complain that they just don't get the leadership buy-in or support to make the changes that the organisation needs. Well now is the right time to develop their influencing skills to help them engage successfully at that senior level.
- **A process approach** – in all things, through people.

Another quote from IRCA illustrates the scale of the problem that demands new rigour from auditors:

4. Reinforced process approach. Although the process approach was introduced as far back as the 2000 version of the Standard, evidence suggests it is still not well understood. Given that the 2015 version of the standard places increased emphasis on the process approach (this being one of three core concepts underpinning the new standard as detailed in the Introduction) it is essential that auditors now understand what is meant by the process approach and what they will need to do in order to assess its implementation by the organisation.

These skills are intrinsically linked and whether it is auditing, consulting, performing an assessment or delivering training, it is essential to combine highly developed change leadership skills with a process approach – you need to have both to be effective and successful.

Many auditors will become improvement facilitators and the other way round. Such a union can only be good for organisations and individuals alike.

Are the changes being met with enthusiasm? Ask any organisation and there are always people who will avoid, systematically oppose or obstruct change. But for many the changes will provide a stimulus for its leaders. They will recognise the opportunity it presents to:

- adopt systems thinking
- process management
- improvement through variability reduction
- integrate their improvement and audit functions
- achieve great new results.

And get their Certification thrown in!!

Assessment

The disciplines of Deming's approach therefore mean that continual assessment and monitoring has to be built into everyday work. This naturally integrates with formal audit, internal or third party, to Management Systems Standards and ISO uses the Deming Cycle as the framework around which all management systems standards, including ISO 9001 are now based. But actually this sequence is all too rare in practice. Just as you would naturally taste a casserole to see if it needs more seasoning, so any manager should keep their eyes on their processes, as they are working – wouldn't they? They tend to look at the spreadsheets rather than going to see how the work is working, mostly only testing outputs, and that's too late, just as it is for a casserole.

The stages of the Deming Cycle help leaders understand their priorities and develop action plans, and also are effective for standardising and improving the work that is performed every day by their people. Through these methods and tools, the whole organisation will be able to demonstrate its engagement in understanding how the work works and how to improve it. It's this approach that makes the real difference to the organisation's performance and thus achieve a successful ISO9001: 2015 registration as a natural consequence, not a special effort.

Our recommendations for managers

There's a lot for anyone to think about in all this. One's principles, methodologies and tools are all potentially challenged by the changes, so there may not be quick answers. Here are some recommendations in getting started quickly. Learning through the Deming cycle will follow.

Start to develop a shared approach to the transition by buying this book for your team and your key supplier managers. Get them to read it and discuss the questions at the end, and engage with them on thinking through and acting on the implications.

The CEO

Take an active leadership role in the change from the start. Be determined to learn, show you are learning – help your people. You will find a ready response, and much more improvement than you can possibly expect from an apparently bureaucratic process.

Top managers

1. Organise a meeting of a cross-section of senior, middle and improvement managers and staff, together with internal and maybe external auditors. Make sure you get a thoroughly competent facilitator who is well used to top management politics, and is skilled in the tools that enable such a multi-level group to share and listen. Ideally from outside your organisation.

The purpose of this meeting is to understand what your people know, about the standard and about the tools and behaviours that will be needed for the transition. For preparation you can get the attendees to consider the questions earlier in this chapter before the meeting, and use them as the basis for the discussions.

2. Find a third-party advisor to help you think through the consequences and work up a plan, if you have not already done so. You and your people will need new knowledge; it will pay for itself many times over. Don't delegate this appointment, it may well be you who needs the most help.

Auditors

1. Develop your knowledge of the revisions; the principles and the details. Consider the implications for what you need to learn, and take immediate steps to do so – not just in the technicalities of the MSS but of improvement and transformation, especially the inter-personal skills.
2. Ask others in the organisation to help you work out your ongoing role, based upon what they know of the revision's requirements (you may need to do the educating) and the history.
3. Think of audit as part of the *gemba* process. Acting on behalf of the organisation leaders you are being asked to go to the real place and look at real work as it happens to learn and feedback on current performance to increase understanding of processes, how they are managed and overall system effectiveness.

Improvement and change managers, process owners

1. Organise discussions with peers, line managers, auditors, other improvement staff to thoroughly analyse what the changes mean, and how you can integrate re-registration with transformation.
2. Make sure you develop/redevelop your own knowledge, and your knowledge of third-party training/consulting firms. Start to develop a revised curriculum for your organisation.

Everyday managers

1. Ask/demand help from whichever of your support staff is responsible for the quality management system (QMS). Discuss what the changes are and how to take advantage of them to help you do a better job.
2. Arrange awareness sessions for yourself, and follow it up with more as you understand your needs better.
3. Take every advantage in your work to link improvement with standardisation. Ensure that each is always linked in your people's minds.

Conclusions

We are very enthusiastic about the changes. For many years, thanks to our use of the System of Profound Knowledge, we have been inspired by our experience of leading the application of system thinking, process management, variability reduction, the scientific method and so on. Now these factors are centre stage, acknowledged as the drivers of the improved results that all desire, and we are delighted.

We strongly agree with the direction of the changes to the Standard and are looking forward to working with like-minded people who wish to wholeheartedly adopt the changes that are required.

1. Where are you now?

Purpose of this chapter

This chapter focuses on Study, or Check, of the current situation. Whatever the individual words mean in isolation, it implies that you accumulate data, opinions, impressions and so on in order to be able to judge the current situation.

This prepares for a comprehensive programme to develop the effectiveness, efficiency and adaptability of your operations within your organisational system, and to be ready for a smooth certification process to ISO 9001:2015.

Think about your attitude to standards and audits

At first sight it may appear strange to assert that a new Standard could stimulate transformation in how organisations are run. After all, management system standards like ISO 9001 are never at the cutting edge of their topic – in this case quality management. All they can do is reflect good practice, based as they are on the consensus view of technical subject matter experts. In the case of TC 176 (the committee that drew up the changes) there are 160 participating countries represented and they bring their own views and experience to the table. Consensus is never easy and when a standard is updated there are groups pushing for dramatic change and others resisting any stretch that might affect their interests.

However, this perceived weakness also provides the Standard's strength. A huge number of well-informed people across the globe have outlined good practice while allowing users to satisfy the requirements through any means they choose. Several aspects of the revised standard are actually very different in emphasis to what has come before – as a result of a great deal of experience of poor practice by the committee members, and we describe these differences later. They are seeking to move on, to get beyond the depressing 'conformance' or box-ticking approach to registration that has been prevalent in all too many organisations, and stimulate leadership, integration, continual improvement and proper handling of risks and opportunities.

As you read this book you may find that the approach we describe is familiar, in which case your organisation should experience few problems in transition to the new standard. If it is not familiar then some quite major changes may be needed if you are to observe the spirit of the revisions, and indeed to get the better results that decent process management produces.

We hope that readers who are looking to thrill their customers will see how and where to invest the time and effort required to deliver processes which create customer value, operate effectively and efficiently across the organisation.

Start by establishing the current situation

Any honest leader will admit to finding it difficult to get the real picture of what is actually happening. People will tell you what they think is going on, but their comments will vary, perhaps wildly. You may find work instructions, flowcharts, customer surveys, data of course, but 'facts' will be in short supply. Hopes, dreams, disappointments, pride, conscience, politics, are just some of the intangibles that will filter the reality. They are inescapable aspects of human behaviour.

In particular, several people are likely to say; 'this is a simple job, but we overcomplicate it, just do this, or stop doing that'. If only that were true. Few jobs really are simple, and many are the ways to overcomplicate even those that are. Let's try to get to the essence of what is happening.

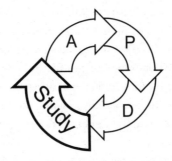

If a tree falls in the forest when no one is around, does it make a sound?
Attributed to George Berkeley, 1710

This ancient riddle has many possible answers (just look it up on Wikipedia!), but in the forest, or maybe more of a jungle, of your work, a few of them are highly relevant. Across your department, and among those influenced by it, and those which influence it, lots of things are going on. Proverbial trees are growing, falling over, being cut down, leaning on others. Some are resisting the weather, others are ready to tumble at the next puff of wind. But who hears them if they do fall? And what sense do they make of the noise if they do?

As with the falling trees, only a small proportion of the everyday events in your function will register with anyone apart from the participants, who may include customers of course. From our experience at the start of our assignments, we see that even apparently tangible data may not be representative of everyday activities, or be accurate. Most events are not being measured. For instance, a call centre manager may be able to see their staff at work, check when they arrived or how long they spent on their break. But how can they possibly know why every order was lost or won, even if they do watch, and every word is recorded? The great majority of work is invisible to the bosses to most intents and purposes. Operations carried out by home workers, night shifts, delivery drivers, technical support, production staff, these are the equivalents to the trees falling – or not – in the forest with no one there. We all know people who behave in one manner when they think they are being observed, another when not, including ourselves, so making sense of it all is not easy. This conundrum led to Dr Deming's emphasis in his System of Profound Knowledge for managers to have an understanding of psychology.

Some hope that an audit of the work and the processes will provide reassurance, and this has been the enduring drive behind the use of the ISO 9001 standard for auditing and certifying management systems. But it has proven impossible to accurately and dispassionately inspect – audit – organisations that are not soundly based upon capable processes and involved staff. All too often the audit has been an exercise in box ticking, reassuring to nobody.

Getting back to the 300-year-old question about trees and noise. For our purposes the answer is, 'yes, every tree must make a sound, but no one can be sure when, how loudly, why and with what consequences, and so on'. So it goes at work. In your area of responsibility you can never be sure about most individual events. Data about events are abstractions, not 'facts', they are imperfect reflections of reality and should only be used in decision-making with great care. Spending time poring over spreadsheets will not show how the work is actually working. You need to learn the flow of how it actually works and thus be more confident about what might be influencing the data.

> Not everything that counts can be counted, and not everything that can be counted counts.
>
> Albert Einstein

Hence you should approach the subject of this chapter: Where are you now? with caution. You can get a feeling for it, good enough to take some decisions about possible changes to test, but don't hope to be certain, and beware of those who assert that they do know. As we have already said, much data that exists is not reliable, and not used very well. Many decisions are taken without good data and little context – we call this 'tampering'.[1] This needs to change – only data that are useful should be collected, and decisions should refer to data in the context of the process that generates it. You will find Chapter 5 of Dr Deming's book *Out of the Crisis*, 'Questions to help managers', very useful. If you struggle to answer

---

1   Dr Deming used to conduct his 'funnel experiment' to demonstrate the inevitability of well-intentioned decisions generating more variation if they are not informed by control charts to interpret the data.

or explain then we recommend asking your quality professional, who should be able to help, or take steps to master them yourself.

Consider your ambitions

ISO Clauses

ISO 9001 also starts with a 'Where are you now?' section under Clause 4: 'Context of the organisation'. This clause introduces many new requirements with the aim that organisations take some time out to decide what it plans to do before launching into documenting its quality management system. Where the organisation has carried out a thoughtful assessment of where it is this insight can be carried forward into further detailed planning and this is covered in subsequent chapters. If you like: a high-level 'Study – Act' before the subsequent 'Plan – Do'.

4.1 Understanding the organisation and its context (Who are we and what do we do?)

Investigation of organisational context is carried out as a matter of course. In clause 4.1 the organisation is expected to look at itself and the market/environment it operates in. The purpose of this review is to enable it to understand what is important to enable it to grow sustainably and profitably by delivering products and services that meet customers' needs while at the same time meeting market expectations and legal requirements. The requirement is to look at internal and external issues and those it needs to consider as influencing or affected by the QMS.

4.2 Understanding the needs and expectations of interested parties (What do customers and others expect of us?)

Current, historic and potential future customers are forms of interested parties and should be the focus of requirements to be met by products and services. Other interested parties provide additional constraints on how products and services must be produced and delivered – including national and local government, non-governmental organisations, landlords, trade associations and even direct competitors. Many organisations spend significant effort on market research into current and future customer needs and include some form of stakeholder mapping as part of strategic planning. All these requirements have to be understood and considered but the organisation chooses which requirements shape the products and services they provide.

This is not a once off exercise but should be regularly reviewed to ensure the information remains current.

4.3 Determining the scope of the quality management system (How big, how bad, how beautiful?)

Again as part of strategic planning the organisation will identify the scope of activity it undertakes. This does not mean it is fixed over time but a scope helps to define the organisation's mission and is used to frame its offering to customers.

All of the work in Clauses 4.1 and 4.2 informs the organisation as to the scope of what it currently does (and could do in the future). Everything the organisation does that could affect the ability to satisfy customers and ensure interested parties are not adversely impacted should be considered as part of the quality management system.

What it means for you

There are a lot of ways to build your impressions, alternative models that suit some people more than others in different situations. One size does not fit all. Try some of them initially, return to the chapter later on and consider others that could be useful too.

A man should never be ashamed to own that he has been in the wrong, which is but saying, in other words, that he wiser today than he was yesterday.

Jonathan Swift, 1711

At the end of the initial study you need to be able to say 'my assessment is that this is how well I think we are doing, this is why, and here are some aspects I am going to concentrate on to improve matters'. You will thus initially be interested in what is happening, how well it works, how problems are responded to, and what the atmosphere is like. You want evidence that indicates the prospects, so that you can develop theories of cause and effect.

Four-student model

In thinking about what kind of evidence on which to base your judgement, consider the prospects for a class of four students. Note, we are interested in their prospects, not just in judgement of their past. They have taken their exams, doing well or not so well, and had followed the teaching process or not, i.e. diligent or neglectful in attending the class. This presents us with four possibilities:

Student	Result	Attended the class?	Comments
A	Passed	Yes, diligently	Good prospects
B	Passed	No, was absent	Lucky this time!
C	Failed	Yes, diligently	Hopeless (student or class!)
D	Failed	No, was absent	Some hope – try attending!

This model was first developed by Professor Noriaki Kano.

Now most of us have seen something of this in our education. Some people sailed through exams without effort, others who studied hard, failed, and vice versa. The examination system is only interested in results: it has done its job if it has separated 'good' from 'bad', regardless of reasons.[2] It does not discriminate between students who learned the

---

2   This is a big topic of course, and illustrative of the overlap of our theme of process management with wider society. Because the task of examinations is on the whole to categorise people for further education or employment the results on their own are not indicative of how students studied. Hence the process of educating students is heavily distorted by efforts to ensure they get the best grades. As process managers we are subject to similar pressures – to get the results demanded – but we can illuminate our prospects for improvement by the question 'by what methods were the results obtained?' Once we have this information, unlike the teachers, we are able to work to make changes to our part of the system, and those other parts that impact us.

methods and can explain them from those who can remember lots of facts but don't have the context that the class provides. Without that context they will probably not pass a different set of questions. Exam results are thus poor indicators of potential in the real world.

Thus many people grow up in an environment that judges them on the numbers (how many exams, of what grade) in narrow subject areas. Then they get to work and it seems to be the same game. Everyone is judged in the context of their department on how much money, how much output, defects, complaints and so on. Not assessed on: 'How predictable is this operation?' or 'What are the underlying factors that influenced the score?' 'What was the department's contribution to the whole organisation's performance?' or 'Are people sharing their knowledge with colleagues?'.

People have told us of audits which are too focused on the result – getting registration at all costs, and not paying attention to how the QMS is actually used in the work. Thus, although the organisation 'passed its exam' student types A and B are confused, and those interested in the future performance of the organisation cannot make a reliable judgement. The 2015 revisions in effect demand that the auditors pay attention both to the tangible evidence of the QMS and how the management use it. Only student type A will succeed in this case, and one can be much more confident about the future performance.

The four-student model thus leads to these questions to keep in mind as you move about the workplace:

- What are people doing, and are they following an explicit process?
- What are the outcomes like, if the process is followed/not followed?

Assessment approach

If I have ever made any valuable discoveries, it has been due more to patient attention, than to any other talent.

Sir Isaac Newton

Our experience is that the great majority of the people you encounter, whether they work in your department or elsewhere, would really like things to work well, and would be happy to join in with efforts to improve it. Time and time again we have encountered people who are entitled to be sceptical, based upon past experience, but who want to contribute. We believe you should assume that people started with this positive motivation, even if they look cynical now, and that this should colour your interpretation of what you see. If you demonstrate a willingness to learn, they may be – eventually – more forthcoming than seemed likely at first.

As you move around your department, bear in mind one of the aspects of system thinking: understanding connections and interdependencies. Every component of your organisational system is connected in some way to the others. But cause and effect are often separated both in time and space, and this can lead to hidden sequences that people are unaware of, and thus may not be able to describe to you. As an outsider you can and must ask the questions that occur to you about why things happen, and keep on going beyond the obvious. Here's the value of an auditor who knows enough about systems thinking to ask deep questions, and not very much about the history and politics of your organisation, and thus not anticipating any particular answers or immediately ascribing supposed causes.

Find shortcuts to observe many everyday activities

Attend routine meetings, conference calls and so on. Watch, listen, experience, then have some discussion. Are people evidently following standard operating procedures? Try to limit your questions at this stage to clarification rather than inviting justification.

How well is the central part of your function working?

The core activities that produce outputs for the benefit of the customer are often called the *gemba*, which translates as the real place, and can be a useful term to use. It is that part of the organisation where the added value work is going on, and which thus determines its effectiveness and reputation. Watch and listen to what is happening. Not for just a few minutes, but maybe hours. Experience a shift changeover. If there is a suitable job to be done, do it. Warehouse handling, answering calls, clearing tables, anything to both experience the work, and to demonstrate your interest and readiness to get your hands dirty. There are TV programmes that feature top executives adopting disguises to work on their shop floor to see what is happening, but in our experience that is completely wrong. It should always be a visible part of a top manager's job to witness the real work. Having practised it ourselves we never found employees holding back on what they said. Far from it, they welcomed the opportunity to speak out, though as with everything else we can only have heard a partial story.

What do they think they are trying to accomplish?

Find out what they think is the purpose of the department/function you are observing. What are they trying to accomplish? We will return to this question later.

What data do people use?

Find out what information people collect, and what happens to it. What information do they get, and what do they do with it? How good is it all? How do they feel about it all? What information is visible at the workplace? How is it used?

Can people tell the difference between a problem and an abnormality in the work?

This is a more sophisticated question than it appears. In the overwhelming majority of cases people treat each problem as if it was something special, when in fact they are happening all the time. Each problem may get lots of attention, which may be necessary in order to placate the customer, but this attention tends to lead to fixes, which undermine the normal running of the process. But if the problems are the result of regular operations they should be studied in the context of the continuing work, not in isolation.

However, an abnormality, something that is truly exceptional, is a learning and improvement opportunity even if it does not cause a problem. It indicates a shift from the regular performance, which might increase if not identified and rectified. We will discuss later how to recognise an abnormality, and how to respond when you find one, but for now it can be helpful to see if anyone recognises the difference.

Anyone who has received Statistical Process Control (SPC) training should be able to explain this readily.

What are the skills and competences?

Find out about what workplace training people have had, what cover there is for sickness, holidays, etc., to ensure consistency. How do people learn what they need to know – is it just by working alongside colleagues, or is there a specific curriculum? If people learn by watching or listening to colleagues the result is an ever-increasing divergence from the original intention and policies.

What is IT's contribution?

You will need to figure out the role of the IT resources. What they do, as opposed to what they claim to do. How did it get to be that way? Who is the key person to liaise with, is there an ally?

How is work and organisational development managed?

If you work in a large organisation you need to discover, if you have not done so already, who is the owner of the processes you encounter in your department. A process owner is the person ultimately accountable for the overall process performance, to ensure it is capable of meeting its purpose. They have the authority to permit changes to it, and thus you will need to work with that person. However, just because there is a process owner identified it does not mean that the role is being properly fulfilled, this needs to be part of your discovery. But they certainly should not be kept in the dark! If there is no process owner the lines of authority for making changes may be obscure and you will need to find out what they may be.

Make sure you repeatedly clarify the reasons that you are doing all this – to learn, not to judge.

Keep a live diary, as much about impressions as specific items. But do note down anything which seems so obviously wrong and easy to change that you can't understand why it is like that. Keep these to yourself, we will return to making apparently obvious improvements in Chapter 3. Some changes that look obvious and easy may turn out not to be, and you should make your first decisions after reflection.

If you are new in your role this is straightforward, if demanding on your time. You should be able to build a reasonably unbiased picture. If you have been around for a while, people will treat you based upon what they think of you, and you will see matters through the filter of your previous knowledge. You may jump to conclusions, explain things away, and be insufficiently open-minded. You will not be able to ask the naive questions that a newcomer does. An external or new person sees different things to the residents. Your internal or external auditors can give you a degree of independent comment. Some companies encourage cross-divisional assessment, and this can be powerful for all parties. In other cases you may wish to employ an external consultancy, and would expect a rapid and dispassionate summary.

Examine your system in its context

This includes customers, suppliers, perhaps regulators, government bodies, consultants and training providers, trading partners, joint ventures and so on. All of these people have potential for insight. Each has a different relationship with your part of the organisation, but in our experience, if approached on a personal level they will more than likely be delighted to help. If you assume that they would like your processes to work better, and ask for their help in doing that, without obligation at this stage, they will most likely say yes.

Be ready for long stories, you may be the first to take this enquiring approach. Use the 'four-student' assessment approach, and take diligent notes.

In particular, build a picture of the customer experience. We will develop a model of categorising customer satisfaction later in this chapter, but meanwhile, get close enough to some of them to sense the degrees of enthusiasm, commitment, resentment, etc. that are present. Make sure you have a good first-hand exposure, and treat surveys with considerable caution. You may yourself have filled in survey forms – it's an enlightening thing to do – and should be well aware that they tell a very partial story depending on their purpose and how they were conducted. If you can find some customers who are considered troublemakers, or to be a bit obsessed, then go and see them. One vocal complainer can be more worth listening to than a dozen satisfied customers. Anyone who complains a lot probably uses the product or service a lot and clearly cares about it if they are registering their point of view.

If you are in the retail field look carefully at the files that may have been passed to your press relations office. How did things get so bad that a customer was driven to that degree of effort to complain? How is it that the PR people can perhaps get things done that your function would not or could not organise? Are there targets or bonuses that conflict with providing customer satisfaction?

Where does the variation come from?

On principle, consistency of outputs is generally a good thing, but hard to maintain. Customers like predictability. You should recognise that variation comes from everywhere. It is as likely to be magnified as ameliorated in your part of the organisation, and variation in your outputs will be causing your customers many problems. It is hence always valuable to take steps that minimise variation, even if it is not clear why at first.

Some degree of variation is inherent in human organisations. Nothing repeats exactly, and all processes have a tendency to become increasingly disordered unless managed otherwise. For instance, people forget their training, machines wear, suppliers change their products, customer expectations move on. Keep on doing what you have always done and this tendency to disorder (known as entropy) will keep on generating ever more variable outputs. The manager's task must therefore not be limited to administrating, trying to keep things the same, they must be a leader of continual change, just in order to avoid deterioration. This is a big subject, and we hope you will be interested to study it further. But for this book, you only need to accept a couple of assertions:

1. The outputs of all processes vary, and you need to know how much variation is tolerable to your customers.
2. It is useful to distinguish between variation that's part of the regular operations, and in itself can't tell us much (common cause), and variation which is surprising (assignable cause), and hence worth noting from the point of view of guiding your decisions.

We will expand on these two assertions in Chapter 3. For the moment let's consider where the variation comes from. A manufacturing version of Table 1 was proposed by one of the early Japanese experts, Dr Genichi Taguchi in the early 1960s, and we at PMI added the service component in the late 1980s. The proportions are of course approximate, but they have been endorsed by thousands over those decades – people who come to recognise that variation reduction demands attention to more than just controlling their own everyday processes.

Table 1. Where does the variation come from?

In manufacturing	In service environments
40% Design of product and process	40% Design of service and process
30% Control of process	30% Control of process
30% Suppliers	30% Customers

Another way of thinking about sources and consequences of variation is to construct a simple diagram to put it into context, as strongly encouraged in ISO 9001:2015.

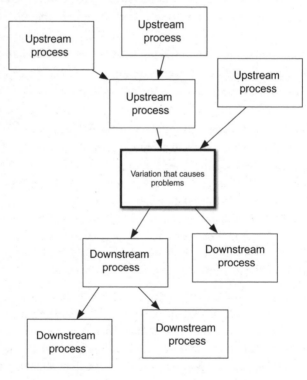

Therefore, if you stay within your department, you can only hope to impact about 30% of the variation experienced by your customers. This is well worth debating with your boss as well as the suppliers. You can hope to make some rapid impacts on how *your* work works, but that may well not be enough. If so you will need wider ranging changes to make the differences your customers will notice, and that means going upstream to design and suppliers. We visit this issue in Chapter 7. Meanwhile, your study needs to uncover prospects for change within your function, so that you can learn and apply within your immediate authority at first.

What kinds of waste can you identify?

Practitioners have found it helpful to think about eight different categories of waste to watch out for. Clearly there are some overlaps and gaps, but if you have these in mind you will find plenty of candidates for attention.

1. **Defects,** which can include missing information, missed deadlines, perhaps incorrect versions or issue numbers of documents.
2. **Overproduction,** such as producing before the output is needed or can be used, thus needing space for storage and maybe finding that requirements have changed. Sometimes people produce more than has been asked for, perhaps just in case of problems, or because they can't actually control the output accurately. We have also seen work being done on non-priority items ahead of more important things, perhaps because it is easier.
3. **Waiting,** or delays between process steps. This can be hard to see, but it is common for waiting time to be much longer than the time spent actually working on the work; many administrative processes are dominated by waiting.
4. **Underutilised talents,** especially of process operators being treated like robots, their opinions and engagement not sought, their experience not recognised.
5. **Transportation waste** is everywhere. Walking with documents, and walking to look for parts on an assembly operation are commonplace. Look for flows to be in straight lines, not zig-zags and loops.
6. **Inventory waste** is linked to waiting and perhaps overproduction. Look for piles of paper, queues of callers (or passengers at an airport), work-in-progress in a manufacturing plant. Inventory is generated by processes that have not had enough thought about getting the pace of the operations aligned with the customer demand.

7. **Motion** is waste resulting from poor design of the workplace. Whether the operation is manufacturing or service it should be laid out to enable the person to complete it with minimum personal effort.

8. **Excessive processing** is to be found everywhere. Too many approvals for expenses, too many reviews for an advertisement are examples. It is often the result of some distant problem, which was 'fixed' by an extra process step, but may even be completely irrelevant now. Look for the value to the customer, encourage people to report this kind of waste. Being new into your job will make it safe for them.

Waste is often invisible to those surrounded by it. They may have noticed it when they started, but having got no response when asking why, it now is just how the work is done. Get into the habit of using this list in discussions, and be sure to respond later if you have decided nothing can be done for the moment. You are looking to reduce people's threshold of tolerance for it, and they need to be rewarded by seeing change happen.

Waste and variation interact, always increasing trouble, never reducing it except by rare good fortune.

Waste is a good subject for attention in process management reviews and quality circles, as discussed later in Chapter 5.

How ready are people for change?

Your exploration is likely to expose a lot of dissatisfaction. Most people will be aware of things not working so well, but have become resigned to it: this is human nature. If they have been putting up with it for a while, or perhaps there have been some change initiatives already that have faded, then they will unburden themselves on any new pair of ears.

However, this may not mean that they personally are ready to change, especially if it involves them taking personal risk, effort or adopting new behaviours. They may think they have tried before. They are also entitled to be sceptical about your commitment and longevity. Being frustrated and blaming others becomes a way of getting through life, and countless historic examples demonstrate it, from restrictive trade union practices, to smoking and drinking, or political beliefs. People may be aware of the consequences of not changing but not accept it applies to them, and will not take the personal steps to move on.

Having sufficient dissatisfaction with how things are is one of three components you will need to mobilise in order to get change on the way. You need to carefully consider the context in which dissatisfaction is expressed. If it seems to be in an environment of complacency, you will need to find ways to confront it, to get people to reflect on the unacceptability of carrying on like this, and put on the pressure as part of your role. However, if it's in an environment of threat and uncertainty, then a confrontational approach is likely to drive people into defensiveness. More effective in this case is for you to be supportive, protective, and inspiring people with a believable vision.

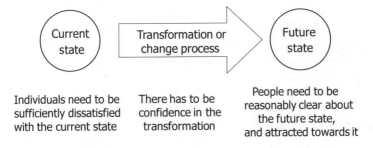

Factors to be addressed in increasing readiness to change

In either case you will need to articulate how things could be if they work properly – the second of the components – a vision. It has to be both attractive and clear, regardless of how ambitious you may be. If there is no defined aim there cannot be a system for improvement.

Recording your findings

You will have gathered a great deal of information, some easily, much rather hidden, particularly that which is cross-functional, that could have the most value. You need to capture it, as suggested earlier, and then to make sense of it, relating and prioritising it to guide your decisions. You need to start at the high level, across your department, before later agreeing with your people a process to work on. We return to system and process mapping in Chapter 2, but meanwhile their disciplines are helpful in note taking.

Summarise processes in their context

Construct a rough diagram to summarise the key components of your part of the system. Under each of the categories list the details such as organisations, people, measures, targets. Note sources of variation and problems. Details of a SIPOC diagram can be found in PMI's handbook *The Process Manager Plus*.

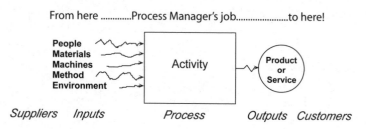

From here ....Process Manager's job......................to here!

People
Materials
Machines   Activity   Product or Service
Method
Environment

Suppliers   Inputs   Process   Outputs   Customers

If you encounter any trained process improvement staff, or Six Sigma or Lean qualified people, they should readily take to this approach and be able to help you formalise and develop it.

Map the process flow

Even if you are not familiar with process mapping, do try to represent your understanding of the work you have been learning about by using some sort of flow diagram. Representing work steps by boxes, and the flow of goods, information or services as arrows, the decisions and feedback as loops, is far more effective than making lists. A flowchart enables you to record questions, observations, concerns and so on next to the point at which they seem to emerge, and in turn this will grow in value as you lead further investigation and improvement.

You could do this mapping work with others in the department, which helps their learning too, of course. Then again, in quickly travelling round on this first visit, it also provides a good basis for your note taking.

You will also find guidance on process mapping in *The Process Manager Plus*, and in online and classroom training.

Develop meaningful goals

However long you take to explore and enquire into the current operations and their prospects, you will encounter a mass of impressions and data, and explanations for them. Some will seem logical, perhaps even common sense, and there may indeed be some obvious decisions to make. However, much of what you have seen and heard will be confused and contradictory, and opinions about urgency may apparently clash with those about importance.

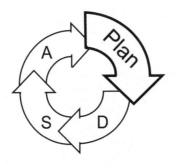

Before you make initial conclusions about actions you need to think hard about the end that everyone is going to need to bear in mind. Then you can attempt a diagnosis of why it's not already being achieved.

This 'Three-Question Model', developed by Tom Nolan, Lloyd Provost, Ron Moen and colleagues, provides a simple reference point throughout your improvement efforts.

What are you trying to accomplish?

This first question is relevant at every scale, from a meeting through to a top management group considering the long term. In many cases you will find that some people have not shared their thoughts about it. It's just yet another meeting to get through. When we introduce this question to management teams it usually turns out that people have assumed they know the answer but actually have contradictory goals in mind. When they have to refer to reference cards you know there's something wrong. Some are focused on the customer, others on costs, others on safety, for instance. It's hardly surprising that they disagree on priorities when it comes to making decisions.

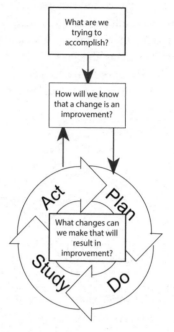

The Standard also incorporates the Deming Cycle as the heart of its learning process, as can be seen from this diagram.

Let's consider first some of the primary goals for a process, and then some ancillary ones. You will need to balance your ambitions across them. You may also find some contradiction between the external goals, derived from the customer, the environment and so on, and internal or political goals from managers. You may have to make a compromise with the internal ones in order to progress with the 'real' ones.

What are the customer goals?

Ultimately the customers are the arbiter of the life or death of your organisation. If they choose others, or simply refuse to engage with you if they find they have no choice, the organisation will cease to function eventually. The following graphic shows the intent of ISO 9001 to assess the degree to which the focus is on the customer.

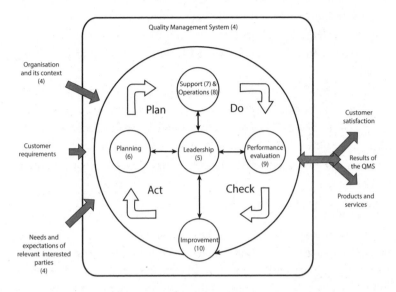

Figure reproduced from ISO 9001 with the permission of ISO at iso.org. Copyright remains with ISO.

This can take a long time to show, as with much of the Western-owned motor industry through the 60s, 70s and 80s. Customers slowly drifted towards Japanese makes in order to find reliability, cost and features that were not being provided by home producers. The home producers misunderstood these reasons, blaming their workers, the exchange rates, appealing to patriotism. Their profits cycled between boom and bust while their Japanese rivals built up cash mountains and global production facilities. The home producers ran campaigns to persuade buyers to act on patriotic instincts, or lobbied governments to protect them, but the tide of better, more reliable and cheaper imported cars eventually swamped them. Rather too late, many companies tried to adopt the principles and tools developed in Japan, and those that succeeded in this are those thriving today. Customers have done well, but many people lost their jobs along the way, and investors had poor returns most of the time.

There can sometimes seem to be rather a lot of process management illustrations from the global motor industry, and the applications can seem distant to a school, hospital or call centre. However, we have all witnessed and benefited from the transformations in the motor industry and the leadership and organisational principles are universal. We can all learn a lot from a global industry that has been learning about optimising itself for 60 years.

One of the key principles is the emphasis on customer satisfaction and delight, pursued over many decades. This has underpinned the financial success and employee satisfaction that can be seen today in the best Japanese-owned companies. None are perfect, little can be copied directly but even their problem resolution methods have useful lessons for all of us.

The Kano model

So, from the first these companies have focused on their customers – how to understand them, satisfy and then delight them. Customers have many ways of reacting to the products or services they experience. The model shown, developed many years ago by Dr Noriaki Kano, provides a very useful way of thinking about what you provide to your customers, and you can use it to categorise your findings so far. It helps find some goals that, if you meet them, customers will notice and appreciate.

The model has three categories:

Expected Expected factors, shown in the bottom right corner, are those the customer assumes will be present without having to ask for them, probably without thinking. As such, they are hardly noticed when they are provided but are a cause of significant annoyance if they are missing. The opportunity thus exists to explore people's dissatisfaction with this focus – what might it be that they are annoyed about but that they don't describe. You may have to observe them using your outputs. What do they complain about among themselves? What bad experiences have they had? What do you notice that's wrong that they don't seem to notice? These investigations often yield a spectrum of information that would normally remain hidden.

Poor performance in the Expected category is thus generally identifiable by watching or listening to customers without prompts. Information resulting from asking questions is in the Wanted category.

Wanted We derive Wanted criteria from two main sources, perhaps being exposed to information that attracts us, or alternatively we may have experiences from our past that we wish to repeat or avoid. There is a simple way to discover these criteria: ask the customer. In particular, it is worth identifying customers who push your product or service to extremes and asking them. You may well find that people who test you to the limit are the best source of information, particularly if they are long-term enthusiasts. Bear in mind the difference between frequent users and demanding users. If you were trying to improve a word processing software package you would get one view from regular secretarial staff, but they are unlikely to expose opportunities for innovation. But if you listen to people trying to use the software to prepare leaflets or posters (perhaps they do not have access to specialist packages), you will see the limitations more clearly and perhaps be able to address them and broaden the appeal of your product.

This search for extreme users is particularly valid with internal customers, many of whom are possibly routinely satisfied or tolerant, but some of whom may have strong views as a result of having to do workarounds, or something non-standard when things change just a little. We do recognise that the dissatisfied and demanding user can be tough to engage with, but their ideas can be well worth the effort.

For an illustration of this we remember a delivery issue with some very large aerospace components. Because they were made in the US but used in the UK, the customer and supplier were thoroughly separated and did not understand each other's facilities. When the suppliers visited the customer they were surprised to find that they turned each large piece over, which required a special crane, before using it. A very short discussion revealed that no one had thought to ask if these pieces could be despatched the other way up. Once the question was posed, the change was easily made. This cost nothing to the supplier and meant a considerable saving for the customer, all down to observation and discussion.

Do not confuse the Wanted category with requests which might emerge from a survey: 'What would you like?' You will get some answers from a survey, but they will be sterile in comparison with experiencing what the customers experience, and asking them on the spot what they think. All of your staff who interact with customers are potential discoverers of customer 'wants', as well as being obvious listening posts for complaints. They may well see your product or service being used for something it was never designed for, and you could be the first to commercialise this if you hear about it.

Delight The Delight factors are at a different level. By definition the customer does not know what they are until they materialise. Their potential may be identified if suppliers develop their understanding of the customer experience and exceed the expressed wants of their customers in a way that they had not expected. Sometimes suppliers can determine delight factors by analysing carefully what the customers do and what they talk about and interpreting this into a new offer. Many successful organisations owe their competitive edge to their ability to generate these delight factors, repeatedly keeping themselves ahead of others by using their specialist knowledge to provide something not previously even thought of. In fact, unless you can keep ahead of what customers say they want, you will be reduced to the level of a commodity supplier, only responding to specifications that are open to everyone. This innovation must be to suit a customer's need or want however; people will not pay for clever technology for its own sake.

Innovation tends to become self-perpetuating if you keep your eyes open. For instance, when Apple first put a movie camera in the iPad they did not expect it to become a major feature. But now that people are using the iPad routinely as their movie camera Apple have been adding refinements to take advantage and encourage its further adoption.

In order to delight people you may consider an alternative to an old cliché; 'treat others as you would be treated yourself'. A customer-focused person will 'treat others as *they* would like to be treated'. Much more powerful, but not necessarily easy, for it implies discovery and empathy – others may not be like you.

A curious feature of this top left part of the Kano diagram is that a new service or product feature does not necessarily have to be perfect when first released in order to generate delight. We remember text messaging for instance. In the first year or so after its provision those telephone companies that provided it won many customers from those that couldn't. Yet a couple of years later the early performance levels had been completely overtaken and it turned out that the launch service wasn't really very good. The company that launched with a just OK product that worked prevailed over the companies that were waiting until the development was perfected.

Today, of course, texting is a 'must have' feature, taken for granted. In general the features you provide to your suppliers drift from top left to bottom right of the diagram, becoming commodities all too quickly. Your long-term survival depends on continually finding things that will delight your customers, allowing yesterday's innovation to become today's 'must have'.

In summary, at this stage you may find that your organisation is not doing well in some aspects of 'must have quality'. If this is the case, such aspects should take a high priority for change. Then you can progress to Wanted or Delight factors. Customers will not be impressed with clever new features if they are being let down on the basics.

What are the financial goals – the voice of the owners?

Customer demand provides the energy for your organisation, but money is its lifeblood, no matter whether you run a commercial business or a not-for-profit. You must be clear about what the financial goals are, who cares about them and how they like to be communicated with. In a big organisation the obvious people are your boss and the financial director.

But you may find that your colleagues have a strong opinion too, perhaps your function has been running over budget and their spending is being held back as a consequence. Or maybe your division's success has been subsidising some poor performers.

In a smaller organisation the financial performance will have a direct effect on all your surrounding stakeholders. You need sufficient income to cover costs, and sufficient cash to reduce any overdrafts. There is no getting away from the need to rapidly become familiar with the credit taken by customers, or allowed by suppliers, and the conditions and costs associated with that credit or debt. Your organisation has to provide for pension and tax payments, and to have a surplus left at the end. This may be for distribution to shareholders, investment in new facilities, or just to build up a cushion against future perils.

Being a process-oriented manager does not, indeed it cannot, change these financial imperatives. Your early investigations will probably have exposed many arbitrary and contradictory targets that may need to be challenged and perhaps changed. But cash in the bank is not arbitrary; it is the seed corn for your future, whether a commercial or a not-for-profit organisation, and you need eventually to be clear on how your work interacts with this number. Money is thus another criterion for prioritising early improvement tests.

This is, of course, a huge subject and must not be trivialised. Arrange to get some training in accounting for non-finance managers, in order to better understand the concepts and language you will encounter. But don't just listen and accept what is taught. The experiences of the last few years entitle you to challenge concepts you don't understand, especially if they seem to run counter to the overall goal of delighting customers as the only route to future prosperity. Surviving the next year in cash terms is important right now, but it must not be allowed to override development and innovation for the future. And anyone who advocates rewarding managers for hitting financial targets has clearly not studied the evidence.

Other goals

Many other goals will become evident during your studies. Some that are known about and discussed; others that you may have realised need to be

achieved but others do not, and are perhaps not even measured yet. Some examples, in no particular order include:
- Pollution
- Safety
- Equal opportunities
- Fair trade
- Carbon emissions
- Employee skills
- Reputation in society.

The above are each big subjects in themselves, with books, training and conferences to inform practitioners. However, the performance in every one of these arenas is the result of processes at work. The variation in your department's achievement of whatever demands are made is the result of variation in the design, inputs and operation of your processes. The methodology to improve any of them should be based upon the Deming/ PDSA Cycle, as described in Chapter 4.

How will you know if a change is improvement?

When you see data, doubt [them]! When you see measurements, doubt them!

Kaoru Ishikawa

This second question of the Three-Question Model follows naturally from defining goals. Every process has a variety of characteristics that indicate its output performance, which are called results measures. Some will be critical, but there are always others that are important. In every case you will need to satisfy yourself that sufficient measures are known that are practical to collect and mean something useful. This is sometimes not as easy as it may appear at first sight.

Consider a pizza. The diner's initial judgement will be based upon some relatively easy-to-measure features such as size, colour and amount of topping. However, their memory of how pleased they are will be dominated by taste, smell and texture, perhaps the stringiness of the cheese, all much harder to put numbers on. You can design in some control of some of these measurable factors, but may have to accept that the chef's value judgement on the day could be more important than, say, a diameter of a pizza or the weight of the dough.

Thus, a task at this early stage in learning about the work is to define some results measures for each aspect of the process considered important. It would be an advantage to talk with some old hands within the organisation, and with some customers, but be prepared to have to modify these views later in the light of experience. Different customers may have contrasting views, and it may take a lot of evidence to build up some patterns and standardise the numbers that will be used. It's quite possible that different patterns will demand quite different approaches. Consider a restaurant providing meals before curtain-up at the theatre. Time is a dominant factor – customers will not come back if they are delayed. But an hour later the diners might be romantic couples, and they may value a lengthy interval between courses. If you wish to serve both markets, all the staff need to be aware of which they are dealing with, especially if an early couple is not going to a theatre, or a late pair wants a quick bite before the next train! None of them will appreciate being interrupted in deep conversation to be asked, 'Is everything OK?' just so you have some data to look at on your spreadsheet!

In summary, results measures are necessary for judging the success of your work. It is also fruitful to clarify results measures for your suppliers' outputs because they may not know which need to be controlled on your behalf. However results measures are often not very useful for monitoring your work as it happens, as they only emerge at the end of the process, and may also be destructive, as in tests of materials such as metals or holidays (or even cake!). Thus you will need to discover internal, or process, measures, that correlate with the results, but are generated as the work is done. We will explore these in Chapter 3. The chef has many process measures, such as the amount of various ingredients, the speed and duration of mixing, the time, temperature and fan speed in the oven and so on.

You need to discover which results measures are most useful during the improvement tests, and then refine them by use in everyday operations. You will realise that targets need to be considered very carefully, and that the data must always be considered in the context of the process. You will also discover that in some cases success is not confirmed by a number, but by a feeling, such as the ambience of a shop, or the atmosphere in a call centre. Data are useful, but rarely tell the whole story.

You need to keep track of the overall process performance through several categories or dimensions:

- Effectiveness: how well they work as seen by the customers and the outside world.
- Efficiency: how much resources they use. Effective processes tend to be efficient as they generate minimal scrap, rework and waste, but apparently efficient processes may not be at all effective.
- Adaptability: how well they can change as needed without losing effectiveness and efficiency. This one may demand generating variety of output, not to be confused with accidental variation of the process or its components.

Your ambition: 'on target with minimum variation'

This can become a universal rallying cry both for the whole organisation and for individual operations. When used with sensible results and process data, in the context of the work processes, you will find it powerful and it can become part of the overall vision for your department.

For 'on target with minimum variation' to be useful it requires that the goals are relevant to the customer. There may also be some agreed criteria that would result in rejection, or a claim, if not met. This could be a time for a delivery, a dimension for a component, or perhaps a reliability target. When you get this right your assessment of the target and variability of the output will enable you to anticipate the customer response.

On target with minimum variation is not the same as complying with a specification: 'OK, Not OK'. In this mindset anything that is apparently in specification is regarded as good, and the same as anything else that complies. Conversely, anything outside the specification is deemed a failure and all failures are equally bad.

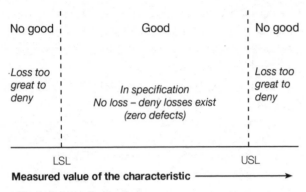

LSL – lower specification limit
USL – upper specification limit

This black and white attitude is often very unhelpful. From examinations at school to inspections of hospital wards or testing of goods delivered, forms are filled in, boxes ticked, claims made against failure. Immense effort is put into getting the right score, regardless of what happens to individual customers, patients, etc. In turn this leads to the loss of valuable insight, because although everything varies, if the only information is pass or fail, there is not much to learn from. Time is spent investigating the few per cent that fail, not the vast majority of outputs from the same system, which also vary but conform, so are not studied.

In addition, the problem can become exaggerated when results are averaged (as shown opposite). Always keep the whole of the data, including its variation, not just pass/fail judgements or averages.

Unpredictable variation leads to costs to the rest of the system, whether they are evident within your part of it or not, perhaps even if they seem to be within the specification. You will find in your investigations that many of your existing specifications are not based upon rigorous analysis and will benefit from extended study and redefinition at some stage.

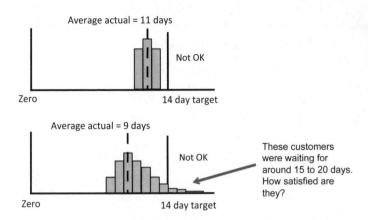

Process capability and Six Sigma targets

There is a potential conflict here. We may desire to get the process on target with minimum variation, but the customer is entitled to require us to manage to their specification. We need some standard way of comparing our achievement to the wider world, in particular with competitors. This is called process capability, abbreviated to Cpk, and is closely related to Six Sigma performance that many readers may have heard of. This is a big topic, too complex to do justice to here. If you are interested, there are many descriptions and training programmes available, but at this stage your basic understanding of processes and their improvement is much more important.

Cpk and Six Sigma relate the customer need or want, i.e. the specification, to the ability of the output of the process to be kept centred over an extended time on the target and within this specification, in spite of its variation.

In manufacturing industry there are tens of thousands of processes across the world that routinely use this approach to reduce failure rates to less than 2 or 3 per million, hence the astonishing reliability of cars, cameras and so on that are constructed in this way. Many such components can be assembled in complex products, without inspection, with the confidence that they will always work properly.

The degree of ambition in performance is for you to work out for your environment. If you have many processes that consist of interactions with other people then failure rates of a few percent may be a big improvement on historical performance. The concept of 'on target with minimum variation' is always valid, and leads to useful discussions.

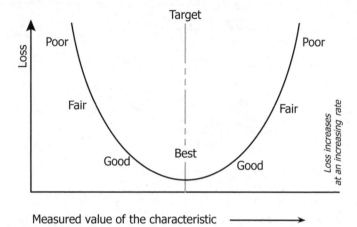

This diagram shows how the variation away from the target leads to increased costs downstream.

Summary

By looking at your organisation in its context you begin to construct the framework for the quality management system and provide foundations for process analysis and management, covered in later chapters.

As a manager you also have a responsibility to understand the circumstances of your current performance, and of the many factors that led to this situation. You may never be able to be sure about either the present or the past, but the better you can relate one to the other the better you can make a diagnosis for change.

So make sure you have paid attention in this study to each part of the System of Profound Knowledge – the organisation as a system, its variation and waste, how it learns, and the cultural and interpersonal realities. You have the basis for moving on to making some changes that will be improvements.

2. Summarise your system, decide on priorities

Purpose of this chapter

In this chapter we cover the 'Act' and some of the 'Plan' of PDSA of what your operations need, and of your revised quality management system. Having made sense of the picture emerging from Chapter 1 you decide on the initial focus.

Simultaneously we also address the requirements for top management covered in Clauses 5, 6 and 7 of ISO 9001. These clauses capture the roles of top managers and development and deployment of strategy. Depending on the results of the current state review in Chapter 1 and any strategy developments, you may end up revisiting processes, variation and controls covered in ISO 9001: 2015 Clause 4.4 and covered in Chapter 3. This is an iterative process and should be periodically reassessed as a source of improvement opportunity.

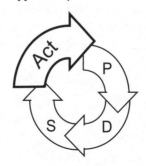

Examine your system in its context

Knowledge is little: to know the context is much, to know the right spot is everything.

Hugo von Hofmannsthal, 1922

You now have a wealth of more or less connected observations and data. If all is well it should be straightforward to summarise in a 'systems' format: an overall picture, the component processes as they are actually operated, and with data and contingency plans well referenced. If this is the case, making the transition to ISO 9001:2015 should be no problem.

It's perhaps more likely that, even if your department is running reasonably well, you have uncovered a lot of small shortcomings. Or perhaps everyday operations are an everyday struggle to keep going, problems often disrupting any semblance of routine.

Whatever the situation, the next step is to get the information organised, and then to prioritise what should be the focus of changes and improvements. Then you are ready to select one process to be a pilot for subsequent work.

We illustrate a series of mapping tools that enable you to understand and represent what's going on, starting with a high-level system overview, developing towards workplace detail.

A system diagram looks at the whole organisation and its context

The first step is to create a map of your part of the organisation in relation to its context. Ideally the leadership team of the whole organisation will have done this and you will be able to create your system as a subset of this whole. But if you are proceeding under your own initiative at this stage you will find that a local map will be very helpful.

This model is based upon that used by Dr W. Edwards Deming from 1950 onwards.

The system map does not replace the organisation diagram, clearly it's important for people to know who their boss is. But the organisation diagram does not show how the work flows, or the relationships between processes, and it does not include customers, suppliers and others, who are part of the system, whether paid as employees or not.

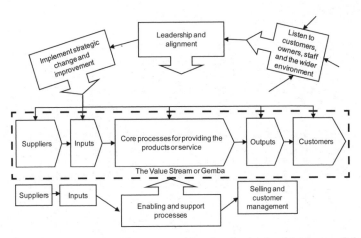

The system map must be put together by the leadership team of the system in question, in order to build a shared understanding of the components, flow, dependencies, purpose, data and so on. There may be some profound differences of opinion, and thus the discussions have an equal value to the finished diagram. Third party facilitation can thus be very helpful to all concerned.

Many issues and problems will emerge as the discussions proceed, and careful records should be kept. It is often the case that priorities will select themselves as the records are reviewed towards the end of the process.

The next example is of a map created by the executive management of a large global organisation. It provided the basis for considerable improvement work, both in the core processes and some of the critical support processes.

ISO 9001:2015 and the supply chain

It's worth considering your wider system as one that specifically includes processes of key suppliers and customers. The success of the service or product in its provision to the customers is dependent upon the process flow across all the components. It's long been the case that big customers have required their suppliers to be registered to ISO 9001, but many have only done so reluctantly and without it changing their practice. The revisions now mean that a joint approach with suppliers can lead to improvements in all the relevant processes, and more standard ways of working and communication, thus saving time and money across the relationship.

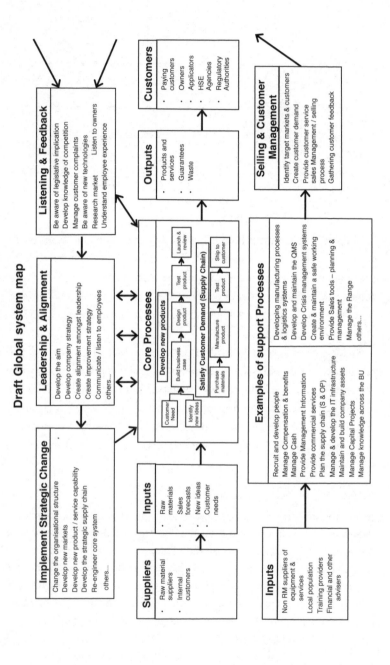

Draft Global system map

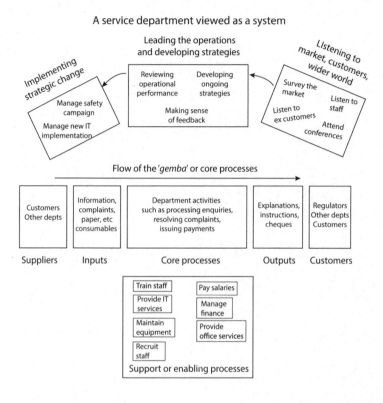

A service department viewed as a system

On the other hand this example illustrates the value of a map at a departmental level. We find that people within the organisation are always fascinated to create and then use such maps. Comments such as, 'now I can see what my job is!' and 'I never saw department X as my customer' are routine.

Create a SIPOC diagram of the core processes

The next step is to summarise the sequences of the core processes, and of various characteristics of each of them. This example is from a pizza restaurant.

Suppliers	Input	Process	Output	Customers
Italia Pasta Azienda Fresh Organics Clients Market Presto Pizza	Dough Vegetables Telephone order Ingredients Kitchen	Name: Order and deliver pizza Purpose: To deliver a pizza to a customer Owner: Luigi Tartufi	Pizza Bill Empty cartons Dirty baking trays	Delivery driver End customer Kitchen staff

Process steps (high level)	Take telephone order	Make pizza	Cook pizza	Hand to delivery driver	Drive to customer	Present pizza and bill	Results measures	Results concerns	Customer needs
Process measures	Number of rings before answering	Time to make pizza	Time to cook pizza	Number of pizzas given to delivery driver	Time taken to drive to customer	Time waiting for customer to open door	On time delivery from phone order to receipt of pizza		Total time within 30 minutes. Hot pizza. Not over or under cooked. Right toppings. Friendly delivery driver with enough change.
Present data	None	None	None	None	None	None			
Goal performance	Within three rings	None	Less than 12 minutes	None	20 minutes	Total time less than 30 minutes		92 late deliveries last month (i.e. 40%). Late delivery means refund of half the money	Date
Sources of variation	Busy in restaurant, not enough staff	Backlog in preparation	Oven temperature	Availability of a driver	Traffic jams, ease of directions	Speed of customer response			
Impact on performance	Time delay, missed phone calls	Time delay	Time delay	Under/over cooked pizzas	Late delivery	Late delivery, cold pizza			Version

Map the process flow

Each process needs to be flowcharted (see overleaf). Representing worksteps by boxes, and the flow of goods, information or services as arrows, the decisions and feedback as loops, is far more effective than making lists. A flowchart enables you to record questions, observations, concerns and so on next to the point at which they seem to emerge, and in turn this will grow in value as you lead further investigation and improvement.

You should do this mapping work together with others in the department, which helps their learning too, of course.

You will also find guidance on process mapping in *The Process Manager Plus*, and in online and classroom training.

Explore the concept and role of process ownership

The role is required for effective management of process-based organisations, but may be absent, and is often not defined clearly. It is even less common for process owners to be trained in the role or for them to be specifically supported by staff skilled in the methods of process management and improvement.

Process owners are responsible for:
- The capability of their process: to be confident that it can consistently, economically and effectively produce the output.
- Response to common causes of variation and the action to be taken on them.
- Enabling people in the process to make changes to the process.

Process owners need to:
- Know what the process is supposed to do and how it fits in the overall system.
- Understand how it works.
- Work with the people in the process to establish common understanding.

A process owner must:
- Be willing to take risks and challenge the 'status quo'.
- Work with others to ensure the process is appropriately designed.
- Be prepared to moderate their own function's desires to prevent sub-optimisation, breaking down barriers where needed.

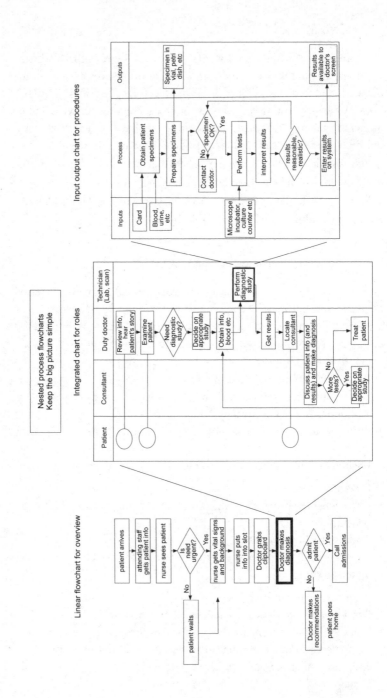

- Devote sufficient time to accomplish their responsibilities.

It is extremely difficult to ensure consistency across a large organisation without this role. Top management need to take it much more seriously in future, and to recognise that, once they have embraced the idea, it takes a lot of training and support for it to become institutionalised. Nonetheless, there are models across many support functions, from accounting protocols to safety guidelines that have ensured consistent operations in their areas of influence. Nothing less is needed if core operations are to become similarly consistent. In ISO 9001 the role of process owner is a delegated one from top management covered in Chapter 2, Clauses 5.1.1 and 5.3 b).

This means that a simple request by an auditor to be introduced to the process owner for a process under consideration is likely to be immediately revealing. If none are forthcoming the auditor is entitled to doubt whether controls are properly maintained across the board. If someone is identified that person will have a very good idea about what's really going on.

Relationship of the content of the Standard to the organisation as a system

The Standard's key clauses apply across the organisation as a system. It is thus clear that using the structure conceived by Dr Deming (described above) will ensure that you have covered all of the requirements in their context, although in the nature of things there are interactions between all of them and almost every part of the system.

ISO Clauses

5.1 Leadership and commitment (How they show they really care about quality)

5.1.1 General

Identifying and communicating purpose remains a top management responsibility as does providing resources to

---

implement a QMS. Individuals within the organization work with delegated authority from the organisation's owners and top management. Hence Deming's comments about 85% of the cause of quality problems lies at the feet of top management.

The buck stops here.

President Harry S. Truman

a) As in the quote from President Truman, the organisation's leaders should accept accountability for the QMS.

b) The organisation's leaders have to demonstrate this through involvement in development and deployment of a quality policy and objectives, again not as a glossy A4 to be hung in the head office foyer but clearly aligned with how the organisation is seen to run and strategies it publishes.

c) They should also be seen to be involved in implementation of the QMS and making sure it becomes part of 'the way we do things round here' rather than a bolt-on set of documents while the organisation continues merrily along its own way.

d) Leaders should also demonstrate that they understand and support the work done in Chapter 3 (Clause 4.4), in particular by aligning resources with delivering process performance and being involved in setting and monitoring key performance indicators relevant to processes delivering what customers want.

e) In ISO Clause 7 later in this chapter there are a series of requirements for resources for the QMS. By ensuring these resources are available top managers demonstrate QMS support.

f) Top leaders should be heard to be supporting the QMS and require their direct reports and their extended teams to work to the requirements of the QMS.

g) Top management should also be visible in reviewing the effectiveness of the QMS. Logically as the team responsible for implementing the QMS (a) – f) above) they should decide whether the QMS is giving them what they require of it. This is covered further in Chapter 5 and Clause 9.3.

h) As an extension of d) and f), top managers should be seen to be encouraging their teams to make sure the QMS works for the organisation – perhaps by releasing them to develop processes, conduct improvement initiatives or releasing them from their 'day jobs' to audit another department. Giving people the space and time to develop and improve processes they work in is one of the clearest indicators of commitment a leader can give.

i) Top managers should be seen to be encouraging improvement.

j) Leaders exist at all levels of an organisation and top managers have a specific responsibility for ensuring that these leaders also act as role models for employees.

5.1.2 Customer focus

> *The purpose of business is to create and keep a customer.*
>
> Peter F. Drucker

For ISO 9001 the prime focus is on customers and, as another leadership role, this responsibility is again laid at the feet of top management, specifically for:

a) Making sure the organisation understands customer requirements including those that relate to how a product or service meets legal requirements.

b) The new requirement in the 2015 edition is for leaders to consider risks and opportunities to customer satisfaction and this requires top managers to lead the organisation in demonstrating that these have been considered.

c) Improving levels of customer satisfaction.

As with all things to do with leadership this does not require top managers to be experts in everything but they have to make sure it happens by putting the right people in place, giving them the tools they need to do their jobs, training and supporting them and giving them time to do it properly.

By focusing on customers, leaders do not allow other aspects of the system to fall down. In their delicate balancing act leaders manage finance, safety, environmental impact, etc. but should always have customers at the centre of strategic plans.

5.2 Policy (where you say what you plan to do)

5.2.1 Developing the quality policy (Leaders develop vision and mission)

A documented quality policy has been required since ISO 9001 was first published in 1987. This requirement remains contentious as it appears to contradict one of Deming's principles to 'Eliminate slogans' but this misses the intent of the requirement and the purpose of a documented policy for quality. Top management has to put together a policy for quality:

a) The process to do this should involve an assessment of the organisation's purpose, context and the market it operates in, as we describe in Chapter 1. Having done that the organisation's leaders are expected to capture what the organisation plans to do to address customer needs in the market it operates in.

b) In the policy or referenced from it leaders should define objectives for quality.

c) Their policy should include a public statement of commitment to satisfying customer requirements and others they see as appropriate.

d) They should also commit to continually improving the QMS.

This policy statement then becomes a commitment standard that leaders can be tested against by customers and other interested parties such as employees.

5.2.2 Communicating the quality policy (Don't hide your light under a bushel)

Having documented their policy for quality as a standard for them to be measured against, leaders have to communicate the content to all employees and other interested parties – again so they can be tested against the commitments in it.

5.3 Organisational roles, responsibilities and authorities (Who does what in the system and how you tell people things they need to know)

All requirements so far in Clause 5 are requirements the organisation's leaders are accountable for and require involvement of top managers as a demonstration of commitment. In Clause 5.4 the requirement is for leaders to manage delegation of authority to others to carry out activities on their behalf. The means of assigning roles and communicating what those roles do under their delegated authority is left free for leaders to decide. Particular requirements that have to be captured and communicated are:

a) The person(s) responsible for making sure the management system works as it should. This may be a single individual or a team leader or responsibility could be shared across regions, functions or by process. It would form part of the role of a process owner.

b) Who ensures processes are delivering what they should? Line managers need to see their responsibilities as including quality and the characteristics (performance, timeliness, etc.) of their work. There will often be a need for inspection or audit sampling activity that could be an internal audit function or perhaps a process owner.

c) Who reports back to the organisation's leaders? Having developed the system top managers need to have confidence it is performing as intended. Similarly the organisation's leaders can delegate responsibility for encouraging people to suggest system improvements and to implement them.

d) Customer focus is identified as a responsibility for top managers in Clause 5.1.2 above but here it is permissible for them to deploy this requirement and delegate to one or more people in the organisation the authority to spread the message on their behalf.

e) As the operating environment evolves so the QMS needs to change but the principles above must still be met so authority for managing changes to the system can be delegated.

6. Planning (How you go about designing your systems)

6.1 Actions to address risks and opportunities

6.1.1 Having done an assessment of current state (in Chapter 1, Clauses 4.1 and 4.2) the organisation has to develop and prioritise plans:

a) Provide confidence the quality management system can deliver products and services that satisfy requirements,
b) Grasp opportunities,
c) Reduce risks, and
d) Improve over time.

6.1.2 These plans need to be built in to the quality management system and its processes including for checks to see the plans are effective.

6.2 Quality objectives and planning to achieve them (How you set a course for where you want to be regarding quality)

As part of planning the organisation has to set and document objectives consistent with its policy (5.3.1) and the environment it operates in (4.1). These quality objectives are generally a subset of the organisation's objectives and relate to how the organisation satisfies customers through the products and services it provides. As part of the quality management system the organisation needs to communicate the objectives (7.3, 7.4) and produce plans (6.2.2) to achieve them including a breakdown of each plan into tasks, resources required and timescales to meet. It remains a top management responsibility to monitor achievement of objectives and manage any changes to plans and the quality objectives (9.3.2).

6.3 Planning of changes

Any changes to plans need to be carried out in a controlled manner.

7. Support (All the people running around to ensure the customer facing teams can do their thing)

7.1 Resources (Give them the tools and they will do their job)

7.1.1 General (another general bit)

Following on from the top managers' responsibility for providing resources (5.1 e) above) the organisation continues on from producing plans (section 6 above) to provide resources to deliver those plans, overarching objectives and to operate the quality management system. As with most decisions there is an 'in house' or 'contract in'/'make' or 'buy' but in all decisions the organisation retains responsibility for effective operation based on internal and external capability. Specifically:

- Does the organisation have the people it needs? (7.1.2 – the bit about people)
- What buildings, equipment and systems are needed to be able to deliver products and services to meet customer needs? (7.1.3 – all the other things you need to do your job)
- Is the working environment suitable for people to be able to consistently deliver products and services to the required quality? (7.1.4 – maintaining the workplace)
- Where, as part of defining the organisation's processes (in Chapter 3, Clause 4.4 c), monitoring and measurement is built in to processes the organisation has to ensure it has the necessary capability to consistently monitor and/or measure with the degree of accuracy required and to maintain equipment suitability.

7.1.5 Monitoring and measuring resources (How you manage your process for observation and your checking equipment)

This section applies principally to the physical parts an organisation produces. It may seem less relevant to monitoring services, but in fact there are many opportunities to apply the thinking and methods to apparently intangible outputs. For instance, an energy supplier needs to bill its customers. This is not the primary output of the system, but the activity may be the most visible part of the relationship with its customers, as the flow of energy itself is taken for granted. So any errors or confusions in billing can dominate the customer perception. You therefore need to treat such outputs as if they were physical products, and some of this section is therefore relevant.

As part of defining the organisational processes in Chapter 3, Clause 4.4 c), monitoring and measuring is built in to processes. Here the organisation has to ensure it has the capability and equipment needed to monitor and measure with the degree of accuracy required and to maintain this capability over time.

You need to decide what measurements are going to be done and what equipment you will need to show the product meets the defined requirements (as required by customers – see Clause 7.2.1). You need to make sure measurements are carried out to meet this plan, and because measurement itself is a process it is necessary to reduce its variation. Therefore measuring equipment must be:

- Calibrated or checked every so often or before use. This must be against standard equipment that has itself been checked with an unbroken trail back to national or international standards. If there are no standards then you need to record how you have checked the equipment.
- Adjusted or re-adjusted (if needed).
- Identified so that you can check if it is calibrated.
- Protected so that it can't be adjusted to ruin measurement results.
- Protected during handling, storage, and any maintenance to prevent damage or losing accuracy.

If you find a piece of equipment doesn't meet specification you need to decide whether previous check results might also be out of specification and record your decision. If you decide you need to, you must take actions for the calibrated equipment and any product you have measured with it. You need to keep a record of calibration and verification (see Clause 7.5.3).

Note: Some other standards may help such as ISO 10012.

7.1.6 Organisational knowledge (the stuff your organisation knows that makes customers come to you)

All organisations depend on application of knowledge to be able to operate processes and deliver products and services that meet customer requirements. Organisations have to make this knowledge available to those operating its processes and to keep the knowledge up to date in the light of changes to the organisation's operating environment. This can include developing or acquiring further knowledge. The organisation has to understand the flow of knowledge into and out of its doors – normally contained in the heads of employees and has to look at how it builds knowledge with the obvious links to personal competence and how it retains knowledge by keeping hold of people and capturing their knowledge in procedures and systems.

7.2 Competence (how you decide people can do what you ask them to do)

All individuals working in the QMS have to be competent to carry out their roles.

Your organisation needs to:

a) Define how you decide a person is capable.
b) Provide training or find some other way to make people capable.
c) Check to see what you have done has worked.
d) Make sure people are aware of how important their work is and how they help satisfy your customer needs.
e) Keep records of education, training, skills and experience (to show they meet capability requirements).

7.3 Awareness (how do people know what you expect). Everyone who works on your behalf needs to know about:

a) Your quality policy (see 5.3 above).
b) Any quality objectives that involve them (see 6.2 above).
c) How their performance can affect how the system delivers products and services and how they can help improve the system.
d) Problems that can arise if they don't follow the system.

Reflect on what you have discovered

As the boss your primary motivation needs to be as a learner and helper, not a judge as must be the case with a third-party assessor. Later we explore the parallels of our approach with that of a medical doctor seeking to diagnose their patient's condition in order to work together to improve it.

Assess your current process maturity

ISO 9001 has been emphasising the need for process management for a long time, but the 2015 revision is more explicit and demanding. This is because the results of any organisation are the outcomes of its processes; success is dependent on leading their operation, coordinating them to deliver their goals. There is no doubt that the long success of the leading Japanese companies in many markets was due to their mastery of all the aspects of process management, and many other organisations have found it works for them too. However, too many organisations have just been going through the motions, processes written up but not representing the way the work is actually done.

Here is the heart of the issue for making the transition to the 2015 version of ISO 9001 pay its way, in fact pay its way multiple times. When you understand the concepts, are proficient in the tools and confident in leading everyone in operations in this way you will get both the results expected and sail through any assessment.

The first step is to establish where you are now: a 'study'. This table provides a quick reference, and a basis for reflection later. You will no doubt find that you can put both positive and negative comments against every level. However, the overriding question; 'is the work delivering to its customer-based goals?' is not in the matrix. No amount of apparently impressive process management boxes ticked can substitute for effective management of the work that the processes enable. Doing the wrong things in a prizewinning way is not what work should be about.

Level	Meaning	Comments at time of first study	
1	The key processes are identified.		
2	Ownership of them has been established, and their purpose is understood.		
3	They are formally flowcharted/ documented and standardised operations can be seen.		
4	Appropriate and visible measures are used to monitor the processes and enable learning.		
5	Feedback from customers, suppliers and other processes is sought and used as the basis for improvement.		
6	An improvement and review mechanism is in place with targets for improvement.		
7	Processes are systematically managed for continual improvement, and learning is shared.		
8	The processes are benchmarked against best practice.		
9	The processes are regularly challenged and re-engineered if required.		
10	The processes are a role model for other organisations.		

In addition, the matrix highlights the simple foundations which must be laid before more sophisticated ideas can be useful. No amount of measuring, reviewing, benchmarking, etc. can mitigate the lack of basic understanding and discipline in process definition and adherence.

Assessing a whole system: The dream of the four 'Es'

The success of the remarkable organisations that have worked to optimise their whole system over many decades can be summed up under four headings, beginning with E. The organisation is achieving its goals, and can demonstrate its management and improvement approach:

- **Everywhere.** Across the whole organisation, including strategy development, everyday work and projects.
- **Everyday.** Leaders understand and can explain the relationship between how they approach their work and the overall improved results they have achieved.
- By **Everyone.** The approach is used in depth where appropriate, and can be explained by line managers and staff routinely, not just the improvement personnel.
- For **Ever.** It has clearly been applied and developed over many years.

If you visit leading organisations across Asia, such as Tata, Samsung, Toyota, Hyundai and Thai Ceramics, you will find these characteristics readily explained by staff ranging from the chief executive to production supervisors. The approach really is their way of working, integrated with their most strategic goals, and operating culture. There are of course many successful Western-owned corporations, but few with such explicit in-depth emphasis on their work processes as well as innovation.

Although the Four Es might seem a very ambitious series of statements, you can make great progress towards them in your department. The case to be built in Chapter 2, 'What should be happening?' will be taking the first steps towards them, and we will revisit the subject in Chapter 7.

How ready are *you* to change how you work?

If you don't change direction you may end up where you are heading.
Lao Tzu

The 'readiness to change' structure introduced earlier is useful in developing your own behaviour as time goes on. You will have found plenty of information that leads you to believe that 'we can't go on like this anymore'. As the leader you will constantly have to think hard about your own readiness to change, and work on it.

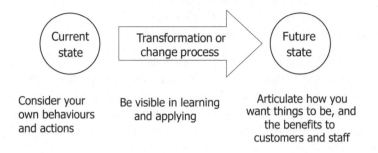

Diagnosing

Much of what you do at this stage of your investigations has commonalities with medical diagnosis. When you visit a medical practitioner with an ailment, they will follow a series of steps to diagnose what is going on, in order to make some intervention to help you get better:

1. They will observe your symptoms, measure various characteristics, ask for information, and so on. They will (should) seek information not just about the apparent immediate problem, but also, hopefully, about all of your person, and perhaps about your family and your home or work life and so on.

2. They will synthesise all this information into an overview, and compare your symptoms with what they know about healthy people like you, and come to some initial judgement about why things are like they are. They should share this with you to see how much you agree.

3. They will then outline a prognosis to you: what is likely to happen if things carry on as they have been. This may well be quite depressing, but if you have understood how they came to that opinion, bearing in mind what has happened so far, you should be ready to do some things differently.

4. The next step is they suggest that if certain changes are made, or treatments undergone, then here is a new prognosis, hopefully an improvement on what is likely if nothing changes. These changes may be as simple as taking some medication, but could well involve you in changes of diet, exercise, smoking, drinking and so on. In other words, it is rarely enough for the patient to be a neutral recipient of a treatment;

their actions are likely to be fundamental to their recovery. As we all know it is unlikely to be easy for the patient to make these changes and persist in them for years. Most people need ongoing support.

Hopefully you will see that your activities so far have been moving between 1 and 3 of these stages working up to 4. You have been listening, watching, measuring, perhaps visiting other reputable organisations to see what success looks like, and talking with your people at every stage to see their reactions to your interpretation. Success at the next stage depends upon them agreeing with your dissatisfaction, and being attracted by your ideas for how it could be, so that they will join with you in trying some improvements.

Part of their preparedness to change is likely to depend upon you demonstrating that you are ready to do so yourself. This may not apply so much if you are new to the role, but may be critical if you have been the boss for a while and are thus looking to improve work that you are already identified with.

Selecting the pilot process

At this stage there will no doubt seem to be too much to do, and not enough time and resources to do it. The temptation to take shortcuts will no doubt emerge. An instruction to the quality manager to 'just get the documentation sorted and ready for the audit while we run the business' has been all too common in the past. No wonder that audit often has a poor reputation.

But the revised Standard is designed to make this approach much harder. There's no requirement to inspect documentation, no template for what the documentation should look like, and no recognition of a 'representative person' to be interviewed. The auditor will be interested in what's actually happening, what the boss is doing to lead the system, how people know if they are following standard processes, indeed are these process capable.

Therefore, a lot of change and improvement is likely to be called for and, if properly led and implemented, it will lead to much more effective processes, more satisfied customers, more fulfilling work for staff, and lower overall costs. Such results have emerged time and again in improvement programmes[1] since the idea first emerged in the West in the 1980s. However, in all too many cases the efforts have not been institutionalised, and entropy takes over, back to the disorganisation that always wins if leaders take their eye off the routine need for standard operations together with systematic change and improvement.

So, the reason for working on one process initially is to learn what is involved in getting it standardised and capable of routinely satisfying customers. In this state it should readily pass an audit. The activity in formalising such a process will demonstrate the work necessary to get the whole of your part of the organisation confident of the evidence of its capability, and ready for audit.

The choice of which process may seem obvious to you, but you need to engage others to ensure their buy-in, and readiness to commit to whatever efforts and changes prove necessary. If it's not obvious it's even more important to get everyone together and agreeing.

Of course, if your investigations so far have shown too many problems to be reasonably confident that standardisation can work well at the moment you will need to select a process for comprehensive attention. This will require an improvement project that will develop its required capability and incorporate the development of standard operations to maintain it. This is covered in Chapter 4.

Either way, taking care to develop consensus among your team is vital.

Agreeing what to change, planning a trial

This is step four of the diagnosis. You need to be clear about what you think is worth doing and why, and ready to build the confidence of your key people to go along with it. Unlike a doctor, it is most unlikely that you will be able to propose a tested, reliable solution – it will have to be discovered, hence the need for careful trial. No one has been in quite the same position that you and your department are in, no one is trying to get to the same place, and no one has trodden the path you are embarking

---

1   Improvement programmes have gone under many names, from TQM to Six Sigma, Lean and so on.

on. This may seem a little intimidating, but is where the principles and methods come in. They have been developed over the decades in similar circumstances, helping leaders to learn how to make the right plans to use the proven tools.

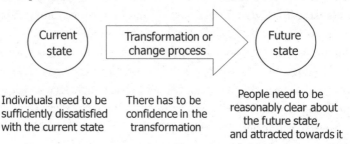

Current state	Transformation or change process	Future state
Individuals need to be sufficiently dissatisfied with the current state	There has to be confidence in the transformation	People need to be reasonably clear about the future state, and attracted towards it

You therefore need to find ways to get your managers, supervisors and staff involved with hearing about your initial conclusions, and jointly agreeing where to make a start. It will require a meeting – the first of many – to ensure you are all understanding the issues and agreeing to the actions.

Diagnosis and planning meeting

Doubt is not a pleasant condition. But certainty is an absurd one.
Voltaire 1694–1778, French writer

This activity is about sharing theories – explanations of cause and effect about the work – built on your study so far and people's prior knowledge. The theories must not be thought of as having to be 'right', but they do need to be made explicit at this stage so that a later study can use the data in the trials to see if they stand up to the real world.

The meeting should carefully follow sensible disciplines to help it work properly.

1 Before (Plan)
- Clarify the need with those attending.
- Circulate the objectives and agenda in advance.
- Make sure the arrangements for the room, refreshments, etc. are organised.
- Get the roles and responsibilities agreed ahead of time.

- If there are any other key contributors, make sure they know what's expected of them.
- Think through possible barriers and provide for them (such as your boss calling for another meeting on that day).

2 During (Do)
- Keep to the agenda.
- Take expectations and hopes from the attendees that may be in addition to the agenda items.
- Honour the timing contract and negotiate any changes with all the attendees.
- Practise and facilitate good meeting behaviours.
- Provide a 'car park' for new issues that haven't got time to be debated.
- Watch out for jargon.
- Identify clear next steps.

3 End of the meeting (Study)
- Review the next steps and ensure someone has agreed to take each one forward.
- Evaluate the effectiveness of the meeting, in regard to objectives and expectations, and capture benefits and concerns.

4 After the meeting (Act)
- Circulate a summary including next steps, promptly.
- Incorporate benefits and concerns into next meeting plan.
- Lead or facilitate the activities on the next steps.

Activities in this first meeting should include reviewing the process map of your function you created earlier.

Select the first process for standardisation with ISO 9001 in mind

With this context in mind you have the best chance to select a part of your system that all would agree is suitable for standardising: with perhaps some problems but thought to be capable of being dealt with in a few weeks.

In this meeting you should also ensure that everyone has a chance to suggest immediate and obvious improvements or changes that don't require policy decisions. There are likely to be many candidates:
- Obvious duplication of effort.
- Collecting data that never get used.

- Replacing broken furniture, damaged signs, perhaps some decoration. Anything that is a result of lethargy or disinterest in the past that people think influences the atmosphere.
- There could also be ideas about workplace organisation that you may want to try changing, but don't expect miracles. Such things are symptoms of other attitudes, as you may remember about your teenage bedroom!

The first meeting provides a vehicle for you to show how you intend things to be – following processes that help people see what is happening, and sharing learning so that they can be improved. There will be many other meetings, and each of them needs to follow this approach. You will refine them with practise, and set the scene for activities, especially meetings, that you cannot see.

Summary

You have now looked top down at your organisation as a system, and agreed on the processes that are the best candidates for developing standard operations that can demonstrate the value of the approach. You have carried out a gap analysis in Chapter 1 covering ISO 9001 Clauses 4.1, 4.2 and 4.3 and leaders have developed objectives and a strategy for quality management in this chapter. These processes are also subject to audit and it has become clearer that your internal auditors and those carrying out assessment for your chosen certification body probably need an additional range of skills from those currently used. In particular soft skills in assessing strategy development and interviewing top managers are key. In the next chapter we start to build standardised processes on these firm foundations and address the detail of ISO 9001's process approach as defined in Clause 4.4.

3. A pilot process: learn how to standardise

Purpose of this chapter

To learn by doing: uncovering the issues involved with standardising one process and informally auditing it to the revised requirements. Preparing for standardising across the board.

> *Cost is more important than quality but quality is the best way to reduce cost.*
>
> **Genichi Taguchi**

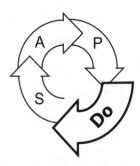

The method used for standardising your pilot process is covered directly by a small number of ISO 9001 requirements but the output directly impacts a significant number of them. This includes the requirement for the organisation to consider risk and opportunities, and to develop processes with these in mind.

The pilot is for learning

It is essential to take a step-by-step approach to all change, using the Deming Cycle to ensure that everyone is clear about the objectives and of the need for openness in comparing what was expected and hoped for with the reality. It may be that your existing operations are acceptable and could gain acceptance in an audit, but that might be a mixed blessing. As we've already explored, change and improvement needs to be continual, just for performance to stand still. Thus, even when things seem OK, you should hope that this first round of standardisation will expose needs for change, and ideas about what to do. You should definitely welcome adverse comments from any audit! Easy to say, hard in reality.

Developing standard operations

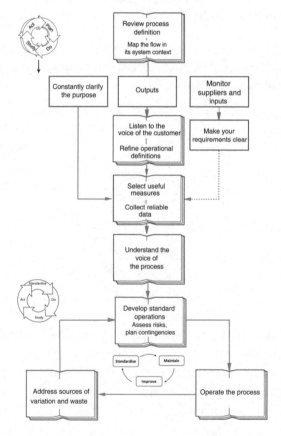

Standardisation is far from a mechanistic process

Deep down, most people acknowledge that logic and evidence, tools and methods are not enough to ensure the success of any change activities. This graphic, which clients often refer to as the hamburger model, illustrates that change and improvement efforts, hopefully moving from the unsatisfactory present to the desired future, have several interacting components.

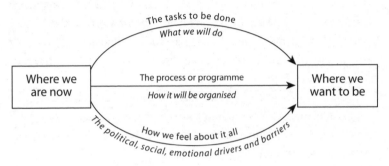

After Jack Gibb

The tangible parts of a change – the tasks and the process – are only part of the picture. Any number of factors can influence people's thinking and behaviour to the exclusion of apparent logic; how they feel about what is going on, or what they imagine might be coming, or why it's like it is at the moment. You only have to watch a sport to see this. Highly competent players have bad days; middle ranking athletes may have a run of exceptional form. Sometimes you wonder if it's the same tennis player in the second set that you were watching five minutes ago in the first one. That's part of the appeal of sport to a spectator; however frustrating it is when your team unaccountably fails to perform as it could. Just to complicate matters, we tend to see patterns in performance where there are none – reading significance into the natural variation that leads people to be better on some days than others for no particular reason at all.

Or consider politics and religion, where people's beliefs may (only may!) influence how they interpret particular events. Some people will favourably react to a private company providing a service, others to a state-owned company or agency, even if they are offering roughly the same services. Religious beliefs may affect people's opinion about a restaurant (pork, beef, vegetarian, etc.) so that they might not make a dispassionate judgement about the competence of the chef. A subsidiary of a large organisation may be regarded with scorn by its fellow subsidiaries. People in big companies tend to look down on those in their small suppliers. All such influences mean that 'facts' to one person may not look like the same facts to others.

Or consider a person's reaction to an airline. If a past experience includes being exceptionally well treated, they are likely to think more favourably about that airline in comparison to someone else who was perhaps refused boarding by the same company for being late. They may carry these feelings for years.

All these factors add up to a set of assumptions that each of us carries with us, influencing every waking moment in ways we usually don't think about. If we make assumptions about others we are likely to come to some very wrong conclusions.

ISO Clauses

In the same way as we recommend the organisation uses a step by step approach to managing change based around the Deming Cycle so ISO encourages users to analyse each process in detail using the requirements in Clause 4.4 below.

4.4 Quality management system and its processes (Why you're reading this)

Having gone through an activity of strategic planning the organisation arranges itself to be able to meet requirements of customers and interested parties. It does that by developing and implementing processes. Sub clauses a) to h) propose a step by step approach to developing those processes implementing them in a robust way and then checking to see they continue to deliver what is required.

This clause is the heart of 9001 and particularly the process approach to quality. In order to 'determine' your processes you have to understand them in detail.

a) ISO defines a process as a set of interrelated or interacting activities that use inputs to deliver an intended result – seems a bit of a 'duh' definition but in the course of looking at your processes from first principles this is a great opportunity to look at what you need before you start and what you will provide the end customer when you're done.

b) In the same way as you can spend an eternity discussing 'which came first, the chicken or the egg?' so you need to critically examine process interdependency:
   - What processes rely on others to operate?
   - What people work across a number of processes?
   - How are resources shared across processes?

c) How do you make sure processes deliver what is expected? Standard ways of working (whisper it: maybe including documented procedures), IT systems, plans and quality checks, supervision needed, key performance indicators. All of these are ways of ensuring consistent process performance covered in Chapter 5. The process you developed earlier, and will refine during the trial can be an alternative method to traditional documented procedures. Later in this chapter we outline how you can capture process flow using a flowchart and how to map a process as an alternative to traditional documented procedures.

d) Following on from b) above where you consider interdependencies of processes including shared resources here you have to ensure there are means to enable resources are available to allow the processes to operate as expected.

e) As a little bit of a lead-in to Chapter 2, Clause 5.3 but specifically for processes here, top managers have to decide who does what and ensure they know. An integrated flowchart of your process can be helpful here, displayed for all to see.

f) So for all processes there are risks and opportunities – covered in Chapter 2, Clause 6.1 but very applicable to ways of working.

This can also impact on c) above as risks and opportunities often lead to checks and balances to ensure risks are managed and opportunities are converted.

g) Like most things with ISO 9001 this activity is not a one-off. You have to look at processes periodically or in the event of a problem to see if they are still fit for intended purpose – and if not make changes necessary to make them work properly.

h) Even if the processes are working well you still have to improve them – and, to be fair this all fits in with the context of the organisation – markets and expectations change and you have to continually improve just to stay still.

The organisation decides what level of documents and records it needs to manage processes and has to maintain them in accordance with requirements in Clause 7.5 below.

Use the pilot to uncover assumptions as well as testing theories

As we have mentioned, we all have prior experiences, culture and assumptions that we don't think about as well perhaps as more obvious ones that we do acknowledge. They all influence our judgement and decisions to some extent. It's part of your job as a leader to use the evidence that arises in the planning and execution of the trial to enable people, including yourself, to reconsider how their assumptions affect their interpretation of theories of cause and effect.

Standardisation and audit are particularly rich fields for negative assumptions. We have all encountered dumb processes that get in the way of doing a flexible job. And we have all been tested at school and know that exams indicate little about a candidate except their ability to perform in exams.

The 2015 revision shows that the writers were well aware of these shortcomings – they have maximised flexibility in the definition of standardisation and documentation, and minimised the 'box ticking'.

Beware: when someone asserts that something is common sense, that's the time to be on the alert. Whose assumptions are in play, whose values, experience, skills make it common sense? In the 1990s it was

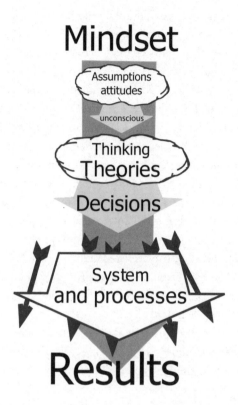

widely thought to be common sense that profitable high street computer retailing was impossible: until Apple started to open their stores and demonstrated otherwise.

Let's take the goal of being 'on target with minimum variation' as an example. We find it is a powerful aid in thinking about the ambition for a process, and in monitoring everyday progress. However, when first proposed to an audience brought up on achieving traditional 'OK/Not OK' targets it can generate a lot of resistance and scepticism. People have probably become accustomed to cutting costs and tolerating or ignoring complaints, and never considered it worthwhile to pay attention to variation in performance if it is within the specification. It may need repeated exploration to appreciate the theories behind the value of looking at all process variation, but at least these theories can be defined. But people are probably not even aware of the assumptions behind the mindset of judging only pass and fail criteria at work. It's just how it's always been, even though they almost certainly don't behave like that in their home life. Both might seem like common sense in their separate worlds, and rationalising them can be surprisingly difficult.

Considering all of these socio-emotional and political factors you might well agree with many of our clients who guess that aspects in the bottom part of the hamburger model might amount to 90% of the reasons for failure in projects. Whatever the numbers in any particular case, it is certainly everyone's experience that an apparently well-planned project, using tools and techniques proven over many decades, can still founder on hard-to-define interpersonal issues or organisational politics.

So it is with ISO 9001. It is difficult to deal in a requirements standard with more subjective topics like communications, motivation and empowerment. ISO 9001 places requirements on the organisation's leaders to ensure awareness of aspects of the QMS and to develop a communications plan but stops short of expectations for how this should look for users of the Standard.

The only way to make progress though, is to actually get on and 'do' something different, following the Deming cycle of course, and then find out about the effect. Thinking and talking do indeed have limits.

Develop your pilot process with the whole hamburger model in mind

For all that it may be impossible to exactly define what went wrong in the socio-emotional aspects of past projects, there are a number of ways you can increase the chance of success for future work.

In Chapter 2 you developed your understanding of the whole system, and of the overall process flow, and this should have led to selecting a process that looks like it can be standardised, able to incorporate any necessary changes without having to stray from your department too much. Ideally it should be capable of changes within days and that a month or so of operation will provide several cycles of operation. In this way everyone can see multiple data points, although five or six sometimes have to suffice to start with. Bigger issues need to wait until you have had some practise, built credibility and understand the barriers as a result of actual work, not just planning or discussion.

Clear your lines with stakeholders

Determine with your team who they consider to be the most important customers, suppliers, allies, and opponents. Agree on how to approach them to explain what you intend to do and why, and to ask for support if needed. Not all will respond positively, such is the nature of scepticism or even cynicism, but in our experience it's better to make the attempt than to ignore them. It's particularly important to engage with the relevant process owner.

Explain that you would like to invite them to some of the meetings as the trials progress. At this stage you probably don't know when that will be, but it's great for both the team and visitors if stakeholders participate in discussions about options, or appreciate success and learning.

Select the right team for the trials and think about facilitation

Depending upon the process, you need a mix of people who know the real work, have technical knowledge, and represent the customers and suppliers. At least some of them need to be volunteers, optimistic and with a good sense of humour. If your organisation has a change agent resource you have no doubt met some of them already, and will know whether you want to use them. Your team meetings will need facilitation, so ideally the change agents should be good at that. Make sure they don't see themselves primarily as technical experts who know best – a sure way to irritate their colleagues and clients. The politics of the work (the bottom part of the hamburger) need facilitation, the details of the methods and technologies in the top part should be chosen after due consideration, not predetermined.

This is a good time to ask an internal auditor, if you have this resource, to come and have an informal look at what you've done so far. If you can get them to conduct a quick audit that would also be helpful, and should expose many issues. Remember that they should be asking new questions, and you may not be happy with the answers – don't shoot the messenger!

Be exemplary with your meeting processes from the start

Clarify at the first meeting what good facilitation actually is, even if some people have been exposed to it.

A good facilitator should:

- Be neutral and objective
- Encourage everyone to participate
- Provide feedback on the team's effectiveness
- Challenge the team to deal with uncomfortable issues
- Provide guidance without directing
- Keep the team on track without controlling, including starting and finishing on time (though that may need authority too)
- Help the team work together to achieve their objectives
- Encourage the development of team ground rules and appropriately use these to help the team move forward
- Assist in reaching consensus without directing
- Listen.

However, don't forget that this is your initiative. The facilitator is there to help you lead your team to achieve the objectives. On occasion you will need to be assertive and perhaps directive, so when you think this is necessary, discuss offline with the facilitator on how to approach it. So although you would like the facilitator to be energetic, don't let them take over.

The first meeting

This sets the scene for the whole trial. This team is going to be interacting a lot, on the job as well as in the meeting room or over conference calls. Hence you need to take them through the approach you intend that they should use, based upon the Three-Question Model.

You should use the meeting disciplines as at the earlier meeting explained in Chapter 2, adapted as you learn more about how they are received, and no doubt to suggestions from the team members. This should then become the required style for meetings, whether you are present or not.

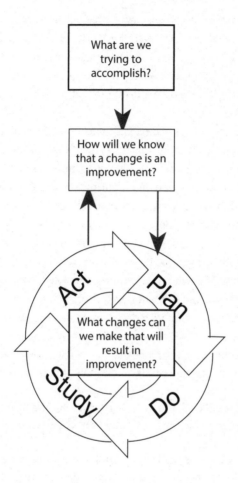

Use appropriate decision-making styles

Different circumstances require different approaches to making decisions. In a fire the fire marshal must direct people, regardless of rank or position, to urgently behave in a particular way. Or there may be policy circumstances, such as employment issues, where you will have to insist on compliance with a particular approach.

In the improvement and change field it is rare for important decisions to be obvious. You may need to select from one of several options that look possible to some of the team but not to others. A quick decision probably means that dissenting views were not heard or that inconvenient data were brushed over. In this pilot work, and even more so during implementations into the everyday operations, you need everyone involved to be supporting the direction taken, even if they have some reservations. Half-hearted support or toleration will not give the best chance of success.

Getting everyone's support, even if some have reservations, is building consensus, and is likely to become increasingly important as complexity mounts. To reach a consensus that really is consensus takes time. People need to feel that they have enough information, and that their voice was heard during the discussion process. This may need more time, away from the meeting, to allow for more data to be gathered, or simply to come to terms with adopting a different course to the one they initially favoured. Allowing this time may not be easy. However, as consensus decision-making has a bit of a bad name with some people, with connotations of endless discussions, and fudge, it is well worth being prepared to take the time, and to be methodical with decision-making processes.

Many of the tools of quality improvement build consensus when used properly, in an open collaborative manner. A few examples:

- Jointly drawing up flowcharts shows how things are, or how people want them to be.
- Displaying honest data in run charts, and talking through it openly, enables people to feel involved, that they understand what the issues are.
- Taking the next step of using control chart limits, which define the difference between noise and signals in the variation, is priceless in understanding whether a change in operation has resulted in a real change in output.
- Generating ideas on Post-it notes by writing silently is a very powerful way of soliciting views that people may not be prepared to speak.
- If an auditor is involved at this stage they need to provide their comments without attributing blame, and to demonstrate that they are here to learn as well.

Taking the time to get everyone on the same page can seem wasted to those who live and breathe the issue being dealt with. You will need to ask for patience, and show it yourself, which is likely to be tough on occasion.

Our experience is that time taken at this stage will be rewarded with less time being wasted later because someone felt left out or overridden. No one can measure the loss of emotional commitment that results from gloomy sessions over drinks, but good consensus decision-making reduces it.

Leading the activities

If changes turn out to be needed before you are comfortable with standardising you'll need to take care. The trials you undertake will almost certainly have to take place in the workplace, possibly with real products or services for real users or customers. It's often not possible to create a laboratory offline. This means that you should ensure the trials are set up carefully, and that provision is made to avoid passing on failures to the customer. Thus you should initially run only one trial at a time, and pay it close attention. Once you have developed a culture of trust in the operation of trials then you can become more ambitious in both numbers and scope, but you will never regret starting at a scale that you can witness.

After planning the tests, making sure that theories and predictions have been made explicit, the following are some approaches and tools that increase the chance of success:

- Use meeting management disciplines, as explained earlier.
- Agree a team frequency, make it public and stick to it. This will determine the rhythm of the work, and hence the overall pace. Having to get something done by the next meeting is a powerful motivator. Make sure the same people attend, seeking the support of your boss to clear time for people from other departments if necessary. (See below for an example of contingency planning.)
- Continually check and increase everyone's readiness to change by sharing your diagnosis, about the unacceptability of the current situation, the attraction and benefits of the ambition, and giving your active support in the risky change activities. Nearly all are risky at first, one of many reasons why you should have only one or two relatively small projects at this stage. Be aware also that your initial diagnosis may not be right, and be prepared not to be defensive if that's the case. Some recent data have shown that medical 'clinicians who were

"completely certain" of their diagnosis ante mortem were shown to be wrong 40% of the time' when tested by post mortem examinations.[1] In problem-solving and improvement we should all be aware of the need to propose theories and then to accept that it's the evidence that should be the deciding factor on whether the theory is useful. Not the strength of argument presented, or the seniority of the person advocating them.

- Take care to keep records from the start on the existing state of the process. These can range from data, written or recorded comments, to videos, photographs, documents, etc. You will be amazed at how quickly you and everyone else forgets just how bad it was, and it undervalues the achievements if you don't capture the starting point.

- Develop a project flowchart for the trial, and a Gantt chart to summarise the timing. These are powerful tools and if you have not already had some training, get some organised for yourself and others.

- Develop flowcharts and other graphic representations of the processes and put them on display, use them as the basis for discussion at the workplace.

- Discover measures that show what's going on in the process, as it runs. Make sure these measure variation if at all possible, not just pass or fail (OK/not OK) data. They need to correlate with important results, so that when they indicate that the process is stable you can be confident that the results will be as expected.

- Collect the data that arise during the trials as it really seems to be. This may be difficult if there has been a culture of targets, budgets, forecasts, possibly with bonuses or appraisals riding on the numbers. If this is the culture there may have been pressure to try to make the data fit the target, and this may have become the default, particularly if it is never talked about or investigated. Many of our clients find that their data are not robust when they start to analyse it for learning and improvement. We have seen measurement variation that turns out to be more than the basic process variation. Some records have an optimistic bias, some pessimistic, some biases will vary from shift to shift and so on. Particularly in call centre environments, data that are entered during the call may be so hard to get that the operator simply fills in a plausible number or reason. The distant analyst at their laptop comes to some conclusions regardless of the validity of the data, so everyone is happy!

---

1   From *Thinking Fast and Slow*, Daniel Kahneman, Penguin Books, 2011, p. 263.

So you need to be aware that a significant shift in attitude to collecting data is likely to be needed for the trials, and this can become quite politicised.

- Plot the data on run charts and, if the expertise is present, on control charts[2] that show how much variation is happening, and how to reliably distinguish signals from noise.

- Inspect all the output if at all possible, so as to protect the customer from surprises.

- Aim to get the process predictable as the first step. Most processes that have not recently been worked on will be affected by multiple factors that make them unpredictable. Eliminating these causes of unpredictability one by one should take priority. Once the process has some predictability you have a better basis for making changes to improve it, and no doubt will be impatient to start doing this.

Until a process is predictable no one can really know if a change is an improvement, a better result may be just luck. Just getting it stable is usually well worth doing on its own at first.

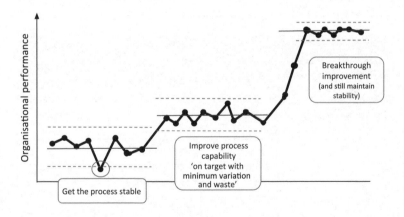

- Are there some immediate improvements to get on with? As mentioned before, these may be obvious things like broken equipment, multiple sign offs, and safety compliance. Make a list of all of them, prioritise by

2   A control chart, or process behaviour chart, represents the natural variation of the process by dotted lines that help determine whether a change should be considered a surprise or not. You should ensure you get practical training to be able to understand and use control charts.

seriousness, urgency, cost and so on, and get on with it. You may also find it worth instigating some organisation activities in the workplace, and there are various 'Lean' tools to help with this. However, you need to recognise that people generally don't make things disorganised on purpose, there's more to it than that. It's relatively easy to have a spring clean, but if the leadership doesn't understand how to maintain it things will slowly revert to the old ways. Ideally you will introduce structured workplace management as part of the changes that improve how the work works, and ensure that daily supervision knows how to keep things up to the new standard. See Chapter 5 for more on implementation.

- Try one change to a process at a time, until you can see the effect. Multiple changes reduce the validity of the conclusions you make. This policy may feel like it is slowing things down, but it is much more robust.

- Analyse risks and build in contingencies. Since there are likely to be several occasions when things can go wrong, you should develop the habit of actively planning for it. This contingency analysis table can be developed in a few minutes, and can save a lot of doubt later.

What could go wrong?	What can we do to prevent it?	How do we know if the preventative step is in place?	What shall we do if it goes wrong anyway?
Key person is off at the time of a trial	Check calendar for holidays, training, etc	Confirm with individuals	Identify a stand-in, ensure they are prepared
	Agree with their boss that they are not to be called away	Note received from boss acknowledging the arrangement	

This is an example of a PDPC (programme decision process chart, one of the seven management and planning tools). The deceptively simple series of questions is the basis for all contingency thinking, and useful in all manner of instances, including weddings and conferences! Where the application is to a safety or performance critical process or product there are more powerful versions such as Failure Mode Effect Analysis (FMEA). These apply weightings of likelihood and criticality to the risks and enable you to prioritise the efforts to prevent or mitigate them.

- Be prepared for surprises to emerge. Building on the metaphor we explored earlier in Chapter 1, in the forest of your department, where trees were falling but no one heard them, during this trial period there will be a lot more people listening! You should be ready for new knowledge to emerge – and this will set a pattern for all your improvement work. The more you learn the more you will discover about aspects of the work that never occurred to you to even think about. Make notes; add such insight to your general diagnosis. In our experience you should be prepared for surprises for up to a year after taking on a new role – many stresses emerge at different points of the natural or business calendar. Friendly old hands can help you to avoid the obvious ones, but you will have to ask.
- Remember also that this is an artificial situation – people behave differently when they have been paid attention to and are aware they are being watched. No chance of a tree falling without being heard in a trial, so to that degree it's a bit special.
- As the trials progress, arrange for your boss to attend to hear summaries and understand progress, barriers and prospects. Take the opportunity to be as clear as possible about costs and use these occasions to gain support for on-going commitments.
- One operation of your newly standardised process is not enough to validate it. You need to put it through much iteration, in different circumstances, looking for evidence of effectiveness, efficiency and adaptability.
- Carry out an audit, using the guidelines in the Standard, and getting help from qualified people as far as possible.

Planning for everyday standard operations

These suggestions are made to help you avoid the trap of making valid changes that do not become part of the everyday work everywhere.

Involve process owners

In your earlier investigations you will have found out whether the processes being dealt with have Process Owners. If there are such people hopefully you have been collaborating with them already, and they will be members of the stakeholder group. There may still be uncertainty as a consequence of the extremely wide variety of interpretation of the role of the Process Owner, and of their profile.

Work with the process management cycle

The process management cycle provides a structure that starts with the design and implementation of standard ways of working, but incorporates continual monitoring and change, so the operation does not fossilise.

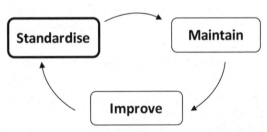

It could be assumed common sense that one should discover the best way to do the work, and then make that the standard, but it is not common at all. One reason is a healthy scepticism about bureaucracy.

Standardisation needs constant protection from association with mindless bureaucracy. The ways of working must not become a meaningless drudge, to be persisted with regardless of the circumstances. That way leads to boredom for the operators, and decline of the performance of the work, for the circumstances of the daily operations do not stay the same forever. All kinds of small changes, from technology problems to customer attitudes, will accumulate to degrade the performance until some kind of crisis finally breaks.

When talking with friends about this aspect of making the work work, their response is often that it only applies to large organisations. However, our experience is that the need for repeatability does not respect the size of your function. In small offices practices can vary widely depending upon who picks up the phone, orders the stationery or responds to a problem. Confusion, disappointment and cost are the consequences in a local business, just as in a multinational.

Develop standard ways of doing the everyday work

We are trying to develop a self-organising system, as far as possible. Good standardisation is an agreed way to do the work, supported by documents that identify both the steps that must be followed precisely and the steps that allow flexibility. It should convey the purpose of the process, its key steps and the customer requirements to bring a customer focus to all work. It should also clarify restrictions and constraints, as well as roles and responsibilities. If it does all these things well, the operators will take to it, and the need for on-the-job supervision can be reduced, to everyone's benefit.

Any improvement project that developed new or modified processes should have worked with operators to define Standard Operating Procedures (SOPs), associated training processes and job descriptions. In principle, the description of the method should be in the form of a flowchart.

Standardisation needs engagement. It must not be seen as a set of rules that result in punishment or reprimands if not followed. Documented procedures created by managers or engineers and imposed on the workforce will probably not work. A procedure manual gathering dust on the shelf is also not good standardisation, and there is also no point going into details on things that do not affect the outcome.

Four rules for standardisation help create a positive atmosphere for its implementation:
1. Standardise only the important factors that impact the product or service.
2. Process operators should understand the 'Why' of the standardisation and be able to influence the process.
3. Managers and technical experts should play a support role to the process operators involved in maintaining and developing the process.

4. Process reviews are for learning and improvement, not reprimand.

Local managers carry the prime responsibility for ensuring that standard operations are made real. One aspect of this is a routine culture of operating according to the standard, not accepting bad inputs, not doing 'bad' work, and not passing on bad outputs, however they are defined and whatever the apparent origins. This is likely to be a radical shift from historic patterns of 'it's good enough, just get on with it'.

An integral part of ensuring the success of standard operations is to ensure that people have the skills and knowledge to carry out the work.

The robustness of standard operations is tested at times of pressure. Questions in case of trouble need to be based on process, not personality. Any manager who urges extra effort in time of trouble is in effect demonstrating lack of belief in the robustness of their operations, and should not be surprised if previous attempts to maintain standard operations are thus undermined. The methods in the Standard Operations Guide should be regarded as being capable of improvement, though only after a disciplined approach to doing so.

We cover the subject of maintaining process performance in Chapter 5, and improving it in Chapter 4.

Display and discuss process performance

The implementation team should work with the process supervisors and operators to develop display boards at the workplace. These should feature the position of the process in the organisation as a system, and express clearly its purpose and contribution to optimising this system. The customers, suppliers and support processes should be identified, and key measures displayed on run charts or, preferably, control charts. Some measures will be those that enumerate the results of this part of the work (this process), others will be indications of performance within the process – analytic data. Ideally the analytic data will show *effectiveness* – how well the process is working in comparison to its design. If these data are as expected the process should be *efficient,* and this will be indicated by the results data. It is very hard to run a process for efficiency, i.e. low cost, if it is not effective. Other relevant information such as training matrices, holiday plans and so on may be added, although care needs to be taken not to overwhelm the space.

A visual display board must be live, being used at least weekly if not daily. It should be the focal point of briefing meetings, and the start point for explaining the process to visitors. Thus it is probably best for it to be done by hand rather than as a computer display, although live information on a screen may be a useful addition.

Keep the IT people on board through the changes

Many organisations have developed IT systems that supposedly determine exactly how the process should be run. It is common for people in and leading these processes to feel that they aren't responsible for their design, and that anyway they can't change anything without prior permission. In some cases of course these same people have developed workarounds that enable them to get results from a system that is flawed. All this can conspire to delay decisions and limit creativity. We suggest you reach out to the IT people, and build a momentum for change, even if their budgets don't allow anything at the moment. The better your developing knowledge, the better position you will be in to escalate later on.

You may not have any changes that immediately affect the IT function, but the chances are that something will need to be different eventually. If you develop a consistent policy of integrating IT people in your process development work you will find that both sides understand each other a little better. Programmers and analysts are often frustrated that line staff and supervisors do not see their work in systems terms – of flow and decisions – and have to make assumptions or guesses as a consequence. The more you and your people understand how the IT people think about work, and vice versa, the better will be any future IT project.

Communicate, communicate (two-way, not just broadcast)

Be sure that you make others who interact with your function aware of what you are doing, not just by broadcasting, but also by dialogue, using language – operational definitions – that they understand. This should include any changes you need from them, benefits for them or the organisation as a whole, and any risks for them.

This communication will become more important as time goes by. You may be managing in a different way to that previously experienced. Or you may appear to be trying an approach which has some history but encountered problems. You need others to be at least neutral, if not enthusiastic, about how you are making the work work better, and this requires them to know about it.

Review

By this stage, from looking around your work to the carrying out of experiments, you have completed one revolution of the Deming cycle. You can see why, in spite of the sequence of its initials, the cycle should start with *Study*. The first *Study* was followed by *Act*: decisions about where to put the first improvement efforts. *Plan* surfaced theories of cause and effect, with predictions of what was hoped for from the trials in terms of process and results data. Without these predictions there would have been nothing to compare with. *Do* was the trials to see how the theories work out in practice.

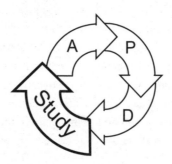

Thus we are brought to the second *Study*, a crunch point in generating learning that can be applied to the benefit of the organisation, both in terms of improving how the work will work, and in how the trials were conducted. A review meeting is needed to confirm the changes that need to be made.

You will need to take care that this review does not get infected with what is probably the default way of reviewing departmental or project activities and data. In the vast majority of performance reviews there is a drive to satisfy some external agency, such as the boss, the parent company or perhaps a regulator. There is probably some target that needs to be hit. In the subsequent ongoing work, numbers that come out just the good side of the target are received with relief, the report can be completed and the meeting closed without too much discussion. Numbers that cannot be massaged to be on the good side will probably lead to recriminations. Much time may be spent adjusting the figures to give the impression of a favourable outcome, or in producing plausible explanations to shift the blame if that's not possible. In either case a depressingly small amount of time is spent on learning about the process or project with a view to improving it.

You might need to make this review explicitly different. It must not be seen as a judgement meeting, an equivalent of a legal trial, with everyone ready to defend their role. It is part of a continual learning process, a second *Study*, to help you and the participants make decisions for improvement. You will find that data that surprised the team is much more useful than numbers which might have come out exactly as predicted. Surprises give the possibility of new knowledge; whereas a trial that ran as predicted means that someone already knew the answer. Or that the data are not so clean after all.

As the trial progresses, and coming up to this decision-making review meeting, it may be hard to prevent outsiders from taking advantage of the growing knowledge, perhaps in pressing for immediate implementation on a broad scale. This is why the preparation described at the start of the project is important, in getting stakeholders such as the boss and the customers to agree to help. At this stage you may need to appeal to them to give you some space and time.

The review meeting

We have suggested that this should be structured separately from the project team meetings. It has a different purpose, which is to come to conclusions and propose or authorise changes in the on-going work. This may require other participants as well as the problem-solving team.

Below are some important factors in conducting a successful review meeting.

- Be explicit about its two purposes:
  - To study the outcomes of the trial, so see how the changes should be taken further, and
  - To study the process of the standardisation work, to see what worked well, what didn't, and what is needed for the future.
- Be exemplary in the meeting management process. You have been using it several times, and it will be familiar to the regular team members. If there are new attendees they need to be taken through the rationale at the start, and to commit to supporting it.
- Get the team to prepare summaries on displays that can be put on the wall, in big writing that can be read from three metres away. You want attendees to stand around them and discuss, not just listen to presentations. Make sure the story is coherent through the displays, so that non-participants can make sense of what was done.

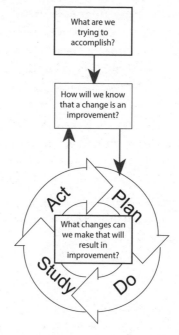

Use the Three-Question Model and perhaps make a presentation to explain it at the opening part of the meeting:

- What you set out to accomplish.
- How you decided you would judge if the changes were improvements.
- What the team intended to do, and what it actually did.
- Display the data in graphic form, using control charts if you can, so that attendees can see how the variation has changed. If it has not changed, it should still be shown, and will influence the decisions you are to make.

Ask the visitors to respond to the displays by:

- Appreciating the work done and the ideas.
- Offering comments on the conclusions.
- Offering their support in making things happen, or asking how they can help.

There are three categories of decision (Acts) that can emerge from the meeting;

- **Adopt.** The results showed an improvement, and the methods of working look practical. Adopt the changes and implement in everyday operations – across all of your processes. This requires an implementation plan, as discussed in Chapter 7. If this is the case, and there is any doubt about your confidence, consider looking for further validation, by rerunning the process with the changes taken away again. If they were useful changes the problem should reappear, if it does not reappear, the initial tests were maybe just lucky, and you still haven't found the cause. If you are a supplier to some automotive manufacturers you may be required to show you have done this validation.
- **Adapt.** The results were unclear: perhaps the informal audit exposed continuing problems; maybe the changes need to involve other parts of the organisation. Adapt the trials and retest with changes.
- **Abandon.** The trials indicate that standardised process does not really work, or perhaps the improvement is too small to justify the effort in changing. This may call for the team to be wound up and starting all over again. Such a decision is of course tough for the leaders, but it's much better to come clean about it and not to carry on meeting and working without any real prospect of improvement. You and the team

members will have learned from the work, and this learning is best applied to the next problem on the list, getting a positive result will be more likely as a consequence.

After the decisions about the project and next steps have been agreed, conduct a proper review of the activities, both in the events in the workplace as the trial progressed, and in the planning and meeting processes. The Four-Student Model outlined in Chapter 1 is the basis. Everyone should be interested in finding out whether if we do this problem-solving work properly it works well, as opposed to being disorganised but it didn't. Or even, if we did it properly but it didn't work well, or were disorganised but it seemed fine. The only secure route for the future is to find an approach that, when you use it properly does a good job for you.

A simple series of questions on each aspect will generate a lot of useful information:

- How did what we actually do differ from what we planned?
- What went well?
- What went not so well?
- What training needs have we identified?
- What would we do differently next time?

Most of these questions relate of course to the socio-emotional factors – the bottom part of the hamburger model.

Having generated the responses, ideally by each attendee writing separately in silence on Post-its, collect them and arrange a separate meeting to digest them and agree actions.

At the close of the meeting carry out a careful action listing, which should include a communication plan:

- What are we going to say to and ask of our people?
- What are we going to say to and ask of our boss and stakeholders?
- What are we going to say to and ask of our colleagues?

When considering who should be involved with making presentations outside the meeting, think about asking a sceptic rather than a known enthusiast. If they agree it may make them think and help their own approach. But not a cynic, probably, that could be too high a risk!

And it should go without saying that the meeting should conclude with a meeting review!

You need to consider what to do with the team, and the hopes and expectations of the individuals. Those who have contributed well are candidates for future projects, and for training in the methods and tools. A comment should go on to their personal records. Look for opportunities for them to present to conferences, either within the company or to trade or management associations. Such recognition will be remembered for a long time, and will bolster you own credibility too.

Extend the standardisation across all your function

Now that the value of standardisation has been demonstrated the task moves on to extend it across the board. However, it still needs to use the Deming Cycle repeatedly, allowing learning to accumulate and mistakes to be caught before they become too serious. This is in contrast with traditional methods and may cause discomfort and resistance. Participants and stakeholders are often used to impressive schedules for rolling out changes and don't recognise the damage this apparent predictability causes. However controlled we would like the implementation to be, natural variation across a human system means it will have to be adapted to some hard-to-predict extent. We have talked at length about PDSA in exploration and testing; now it shows its strengths for implementation projects.

Implementing standardisation is uncertain

Successful military leaders have learned that they need to understand what the goals have to be, to prepare resources and a series of steps to get there, and to allow for contingencies when things go wrong. But they also know they must expect to be surprised because they can't be sure what will go wrong, and to be ready to learn from the experiences, adjusting tomorrow's activities as needed. Managers of organisational change should follow this example, and beware of simple formulae. Predicting the future is always an approximate skill and the best way to deal with surprises is to expect some and have agreed ways to acknowledge them and respond, not to proceed in blind hope it will all be OK.

You will almost certainly have come across projects or programmes that talk of 'roadmaps', 'cascades', or 'rollouts'. The Six Sigma model (discussed in Chapter 4) has two stages named 'Improve' followed by 'Control'. Such language makes implementation seem rather simple, as if hard work, or top management pressure alone will make things happen. Much project management training seems dominated by research and planning of the project, in particular about 'critical paths' and 'work packages'. It tends to be light on learning during the project. Why shouldn't your implementation be similarly straightforward?

The answer lies in the nature of change leadership in comparison with management of tangible, physical projects. A civil engineering project, for instance, is rightly dominated by the imperatives of achieving the target on time, on cost, and with many other defined requirements, such as safety and environmental issues. If you read about a major civil engineering project you will be able to find predictions for when, say, a machine will complete its tunnelling work, years in advance. There will of course be uncertainties on the way, but with a highly competent leadership team you can be pretty confident they will adjust to surprises and force the work on. They will have contingency funds to reallocate resources, and clarity of definition in the goal that means that everyone knows their role. In addition, the managers are specialists in planning and controlling resources to a timeline, so although they may meet new problems they have the benefit of experience in having overcome such barriers before.

Organisational development work, for this is what standardisation is, is different. It can never be about forcing change. It's primarily about helping people to be ready to change how they behave. Anyone can instruct, but no- one has to follow unless they want to. You and your colleagues have to find out what is the best way to increase this readiness to change in every circumstance you encounter. Your goals are probably less tangible than a construction project, and their achievement is unlikely to be made more likely by more pressure, indeed pressure can be counterproductive. This is not to say that a sense of urgency is not important, it's to assert that you are responsible for leading the change, being ahead of the pack, experiencing the barriers and discovering how to get them lifted. Then you can hope that people will follow your example.

Acceptance of a degree of uncertainty is therefore critical to the success of this and every other change that you lead. Nobody has done the task you are about to undertake before. No one has been exactly where you are; nobody has ever been trying to get exactly to the goals you have agreed. There is no plan that you can copy and paste.

In planning your implementation you can and should do a thorough preparation job, clarifying the goals, estimating timings, resource needs, training and so on. You should ask others about their experiences, discuss possible problems and develop contingency plans to deal with them. But you should always expect the unexpected, seek to detect it as soon as possible, make it visible, and rotate the Deming cycle again and again as the discipline for learning on the way.

Implementing standardisation across multiple processes

Meetings are always about increasing readiness to change

The implementation team must contain the line managers and supervisors of the area affected by the changes. This should not be fudged, no matter how much they may claim they haven't got time for meetings. At this stage, if the work is within your own department, you can insist on this. Anyone who suggests they send a deputy for a meeting that is to plan changes in their function is indicating that they don't think it is important. Their staff will take the lead from the boss, and will quickly interpret any tendency to downplay any priorities.

Meeting management processes should be exemplary (again, are you surprised?). As with the exploration, problem solving and review stages, the meetings are the one component that you can definitively lead and ensure they are conducted as you wish. All of the guidelines outlined in Chapter 2 remain valid, with a particular emphasis on time management, as many of the delegates will necessarily be called off their regular jobs, and will not appreciate being delayed. You are likely to find this meeting discipline tested. Some line managers have so far been able to observe your investigation and planning work from a distance, and hope that if they keep their head down it will all go away. As they find themselves pinned down, asked for commitments or even asked to behave differently, they will come up with many ways to prevaricate. Hopefully your problem-solving team contained attendees from the areas being affected by the changes, and they will be ambassadors for the solutions, but that cannot always be the case.

Each meeting will therefore need to contain elements of increasing readiness to change. The emphasis will depend upon attendees' previous exposure. In the first place you must build their awareness of the degree of dissatisfaction about the current situation: you cannot assume they have accepted or even heard of the previous work you have done.

Next is to clarify with them the goals that were developed in the review stage of the problem-solving work, and to refine them with this audience who are closest to the routine work ahead.

The third component is to build the change process with them; the precise steps that are intended to ensure that the new ways of working have the best possible chance of being reliably and quickly adopted.

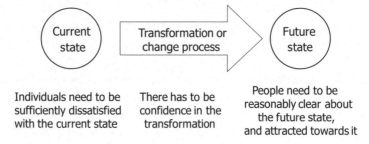

You will find the Three-Question Model is once again a very useful basis for planning the introduction of the changes into everyday operations as shown in this rather basic illustration.

Three-question model

1. To reduce the chance of errors arising from customer queries, and to deal more effectively with any that are not properly resolved.
2. Counting the numbers of errors or complaints. Keeping track of the costs of the change.
3. Change the work instructions, train the operators, inform others affected, and collect data.

Test theories of 'roll-out'

You should test the implementation using the Deming cycle, just as you tested options for the pilot standardisation. The value of tests of implementations lies in their exposure of deeper and deeper issues each time they are run. If you have many people involved in the work, perhaps on shifts, in different locations, perhaps some are employed by third parties, or there is different service or product groups, then there will need to be adjustments to some extent. Only by using the Deming cycle will these variations be regarded as planned learning rather than surfacing eventually as apparent failures.

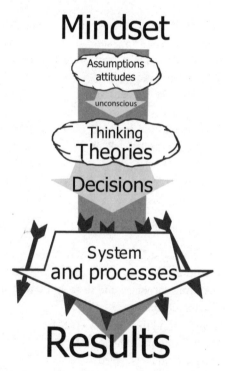

As this diagram shows, behind the work processes operating every day lie decisions taken months or years ago. If they were taken by people who are no longer with the organisation there may be little resistance to doing things differently. But if some of today's managers were part of the original decisions they may still feel strongly about how things should be done. Such opposition may not emerge visibly, it will need to be teased out and people's feelings recognised and responded to.

Multiple operations of the Deming cycle conducted rapidly will flush out both positive and negative feedback each time, until eventually you can hope to surface why things are as they are. There are similarities with your own fitness regime, whose roots probably lay many years back and can take a lot of work to dig out and discard.

Auditing

In this chapter we have outlined a change management process. Auditing this activity is very different from auditing an organisation at steady state. An understanding of change and project management will help. Audit can identify good and best practice to influence roll out of process optimisation. Auditors have to assess the change management process to ensure the principles covered in this chapter are understood and implemented; that our selected trial process meets the requirements of Clause 4.4 and, most importantly, delivers outputs as expected, efficiently and effectively.

Summary

The development of standard operations has been neglected in many organisations. Attention tends to go to those who are creative, who sort things out, fix things or wheel and deal, and few of such people are interested in on-going reliability.

It's worth bearing in mind the four key enablers:
1. Standardise only the important factors.
2. Process operators should be engaged.
3. Managers and technical experts should support rather than dictate.
4. Process reviews are for learning and improvement.

Any organisation seeking to endure is also dependent upon people who create and run capable, responsive processes. The revisions to ISO 9001 favour managers who are ready, willing and able to integrate change with standard operations. But as anyone who has taken part in behavioural profile sessions knows, it's a rare person who likes and is good at both of these activities. Your job as a leader is to make sure that your teams contain people from both fields, and appreciate their different contributions at the appropriate stage of the work.

4. Develop step change and standardise

Purpose of this chapter

To enable you to systematically redevelop processes that are incapable and thus not ready for standardising.

Having demonstrated that a pilot process can be standardised the organisation should look to extend this approach out across all processes. However, some of them will require major changes, and have impacts to some extent across the system. There are no additional ISO 9001 requirements for large-scale projects but many processes have specific requirements for customer, design, or outsourcing for example. The widespread adoption of process improvement creates additional considerations for top management in terms of resources required to extend the approach across the organisation.

Leading larger scale change (multiple operations of the Deming cycle)

If you and your team concluded that none of your processes are stable and capable of meeting customer requirements you will have appreciated that it's a serious decision to postpone standardisation efforts at this time. It's worth revisiting one or two tolerable processes to see if standardisation might generate improved stability because that can be done reasonably quickly and you will learn a lot. This is represented in the left-hand box of this chart.

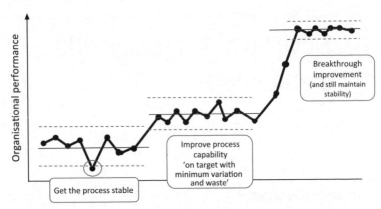

Nonetheless, sooner or later the need for major process change will arise, maybe there's too much variation in how things are done, or if the outputs are too hit and miss. This is where a systematic, repeatable, strategic project approach is needed. It is possible that you can run such a project within your departmental boundaries, in which case many of the cultural and resource issues are simplified. However, it is in the nature of systemic leadership that their interactions cross the system – cause and effect are often separate in time and space – and the approach described in this chapter provides methods for optimising them.

You thus need to lead a demanding project to redevelop a major process to achieve a step change in its performance. This practical experience of leading a large-scale project is priceless if you are later asked to act as sponsor to one; a key role for top managers. The System of Profound Knowledge applies, and the Deming cycle must form the basis of understanding the current situation, and developing and testing theories. Any of the tools so far used will also prove relevant and you will find that tools learned in larger-scale projects will be found to have application in the everyday work. You will be beginning to appreciate that the approach works in all three of the typical roles of a manager: everyday management, project leadership and general senior management. The principles are universal, the methodology needs be selected for the job in hand, and the tools selected from a common pool for the particular application. The details of the tools and their relationship to the methodologies can be found in *The Process Manager Plus*

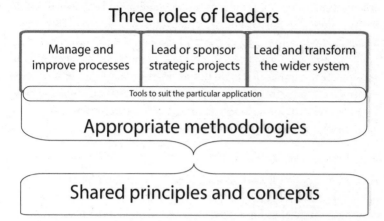

Three roles of leaders

Manage and improve processes	Lead or sponsor strategic projects	Lead and transform the wider system

Tools to suit the particular application

Appropriate methodologies

Shared principles and concepts

Large-scale improvement work does of course need some new disciplines to those demanded so far, in deciding what to work on, how the project work should work, and how its progress will be reviewed.

Which major projects?

In Chapter 2 you and your team identified the priority process for attention, but it's helpful to be aware of the more general issues as time goes on.

The focus for major improvement projects may emerge from the regular work, or be deployed from the top. Senior management must lead the prioritisation work, as there are always too many candidates, and never enough resources to improve everything. Projects need to be focused on objectives that senior management consider important and thus worth the undoubted trouble. In addition, since the implementation of any major project is likely to be time consuming for line managers, a degree of realism from the top is needed to avoid swamping them with too many changes at once.

Three aspects of the organisation generate candidates for major projects:

1. Candidates emerging from the daily work

You can use the Process Management Cycle (Chapter 5) and the Problem-Driven Improvement approach (Chapter 6) to deal with issues as they arise, while identifying the issues too demanding to be handled within the function and thus need escalating. This will depend upon many factors, including the scale of the organisation, the resources and skills available, and the urgency and importance of the issue. You may need to accumulate a number of candidates during the year in order to discuss priorities at the annual planning meetings. Some may emerge with urgency and require reprioritisation. Beware of becoming completely fixed on an annual plan, no matter how times change.

2. Local policies and priorities

Your peer group will also be reviewing its experience of the everyday operations, and hopefully engaging with each other on the market, customers, suppliers, and stakeholders and so on in order to assess future needs. All this information needs to be synthesised into a high-level priority list, and a few major projects agreed.

3. Top management or group priorities

The board of directors, or equivalent governing body, should be continually revisiting the strategic prospects for the organisation. Their decisions will generate needs for change or improvement work that involves large parts of the organisation, and possibly outsiders too, such as suppliers. This is captured as requirements for top management in Chapter 1, Clause 4 and continues with leadership; planning and objectives setting, and control of resources covered in Chapter 2, Clauses 5, 6 and 7.

Agreeing the overall balance of projects

These three categories of projects need to be brought together to form a complete list of candidates. The top management team has to decide which ones are to be given a high profile and support, and which are left to the local management to conduct within their own resources.

It's likely that the revisions to ISO 9001:2015 will generate some direction from the top, possibly a major campaign, but a benefit of applying the new standard properly is that the change activities we describe will be leading to improvements in operations and hence results, not just audit and paperwork.

Coordinating multiple priorities

But: initially the number of major improvement projects must be small

At first there must only be a few major cross-functional projects, certainly single figures and perhaps initially only two or three. We have no illusions about it being easy to stick with this limit, but are also convinced of the need for determination to concentrate the efforts, for several reasons:

• Leading a major cross-functional project is difficult, perhaps the toughest challenge any leader will face until they reach the Board. This will not surprise the reader of this book, as the difficulties of increasing readiness to change are hard enough for a line boss within their own department. These difficulties are harder still for project leaders who will not have executive authority to apply across the functions. Important projects therefore need the best people to be leading them, probably full time, and that is hard for the organisation to arrange.

• The skills needed to lead or facilitate major improvement projects are usually new, and it takes time and practise to develop the necessary confidence.

• Major projects need confident sponsorship from top management. There needs to be a top manager who is ready, willing and able to sponsor each major project. Sponsorship is a non-executive role that supports the leader, provides a connection for the team to the top management, and is prepared to deal with top management politics. Projects that do not get this commitment must be postponed or given a lower profile, for they will not achieve their goal in the absence of proper sponsorship.

- Major projects demand the time and effort of everyday managers and many others in cooperating with the learning and implementation work, and perhaps in changing their own behaviour. Implementing multiple changes often falls short of ambitions through conflict of interest, time and priorities across departments.
- As well as a highly competent leader, major projects require confident and experienced facilitation from people who have in-depth knowledge of the principles, methodologies and the tools. Ideally, these facilitators are internal people, but they must be potential high flyers, and will take many months to develop the basic skills, hence limiting the scale of what can be achieved. Alternatively they may be external people, and are still likely to be in short supply. Recruiting outsiders as facilitators can seem an economical option, but you need to take great care that they have the wide experience and interpersonal skills needed to be able to adjust rapidly to their new environment, and their qualification level may not be a good indication of that.

There is also a need to provide multiple learning projects for those undergoing training to practise upon. They need to be limited in scope, and focused on the use of the methods rather than the delivery of the results. Many change programmes have compromised this purpose by launching multiple high profile projects, with the result that most of the trainees are under too much pressure to deliver, and the projects founder.

This is all likely to be depressing as it seems to limit one's ambition, but we are sure that experienced readers will agree with it. We have seen presentations of programmes where a large number of projects seemed to be generating good results, but in every case that we have investigated the story was not so clear in reality. Some projects had achieved much, but the rest were in truth falling well short of the claims. While public talks get good initial publicity, unfortunately it's often the failures that get the attention over the longer term and give the approach its reputation.

Taking these factors into account, keeping to a small number of major projects that are well run and supported and deliver their goals, is a powerful way of building foundations for long-term change and continuing transformation.

Large-scale projects as vehicles for organisational transformation

We discussed the dream of the Four Es in Chapter 2. 'The organisation is achieving its goals, and can demonstrate its management and improvement approach Everywhere, Everyday, by Everyone, for Ever.'

This table indicates how large-scale projects contribute to this ambition.

Characteristic	Contribution of major projects
Everywhere. Across the whole organisation, including strategy development, everyday work and projects.	By being multi-function, with strategic goals, implemented into everyday processes
Everyday. Leaders understand and can explain the relationship between how they approach their work and the overall improved results they have achieved.	In order for the changes to be permanent leaders have had to make links between methods behind the achievements of the projects and their implementation into daily management.
By **Everyone.** The approach is used in depth where appropriate, and can be explained by line managers and staff routinely, not just the improvement personnel.	Active sponsorship by top management ensures they understand that the approach is different. Continual review of the improvement process means that all involved have helped adapt the approach and don't have to rely on specialists.
For **Ever.** It has clearly been applied and developed over many years.	As the large-scale projects take root, and the approach is used again and again, participation becomes part of everyone's career development.

Leading improving projects, part 1

The hamburger diagram is relevant once more. One of your tasks in leading a project is to increase readiness to change in both the team members and those around the work, in order to minimise the chances of the socio-emotional issues derailing it.

Establishing the team and the charter

The terms of reference of the project need to first be established between the sponsor and the team leader, and should include at least the following:

- A clear purpose clearly linked to performance benefits for the customer if at all possible.
- A list of benefits and of specific deliverables as far as can be understood at this stage.
- A list of team members, and the degree of their commitment.
- An estimated timescale.
- The scope of the project across the organisation.
- Enablers that will help the project succeed.
- Barriers, which may get in the way if not addressed.

This first charter then needs to be worked through by the team, with many of these factors almost certainly needing to be refined by dialogue with the sponsor as the project moves forwards and the knowledge develops.

Improvement project structure

PMI's Improvement Cycle is built on many decades of practical experience. It expands upon the Six Sigma DMAIC (Define, Measure, Analyse, Improve, Control) problem-solving model and can form the basis of a 'Lean' project. It balances the need for individual steps that are seen as achievable with a visible overview of the tasks being undertaken. It can be applied to small and large-scale projects with equal levels of usefulness and rigour and can be easily adapted to innovation as to improvement.

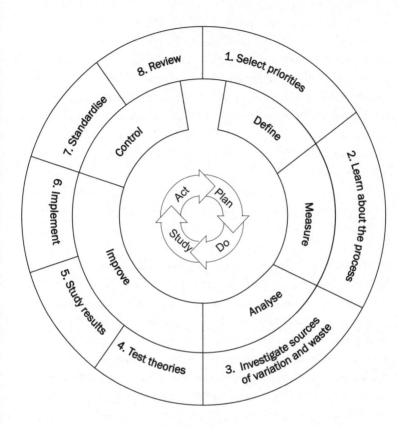

Its power comes from its coherent structure and its strong relationship with the Deming cycle. It can be applied to all types of project and time frames, not only problem-solving projects. The Deming cycle should be spinning many times, helping to develop and adapt theories within each of eight stages. It also generates potential new learning about projects in general and this in turn provides a consistent approach that encourages people to share learning about the methodology across many projects.

Throughout the project you must ensure that meetings are properly run, and will of course find that the meeting management processes you have used so far are valuable here too.

1: Select priorities emphasises the importance of creating a firm foundation for starting the project. It should be launched with a charter, as described above, which ensures that you are tackling the issues in a way that maximises the chance of success. These are of course unlikely to be obvious. The team must repeatedly ask, 'What are we trying to accomplish?' and 'How will we know that a change is an improvement?'. If the whole team is engaged in this process, then the common understanding of purpose pays dividends as the project moves forward. During this phase, the team and the senior management sponsor should engage in dialogue to convert the initial deliverables of the project into logical goals with timescales based on knowledge. This should be formalised into a contract.

The leader and facilitator must beware at this stage of being driven into solutions and implementation before the causal chain has been understood. The pressure from top management may be extreme and the urgency apparent in the style of reviews with the sponsor; hence the need for strong characters in both sponsor and leader roles. They should have had experience at a smaller scale if not previously on such a high-profile project. There is no point using a sophisticated learning structure, intended to uncover new knowledge, if the leadership thinks it knows the answer and is determined to force a solution. A dialogue that avoids wasted effort in further investigation will in this case be called for. You may be somewhat discouraged to read this in step 1 of an 8-step model, but we do counsel such honesty now as being a lot less painful in the long run. Skilled coaching of both sponsor and leader will pay dividends.

2: Learn about the process challenges the current organisational view of what is happening. You already know that the reality of the workplace often diverges hugely from theories that may be strongly held. So it is extremely important for the team to understand the processes as they actually operate. To be certain of this understanding, they should listen to and observe the stakeholder groups involved with the processes. This will reveal what is really going on and begin the process of engaging others in the whole improvement process. At this stage you may uncover some disconcerting assumptions that need careful handling.

It is essential to review the QMS, and indeed it is best if an internal auditor can be a member of the team. Since the context for this use of the Improvement Cycle was the difficulty of operating a capable, standardised process you would expect the QMS to be poor.

This phase is concluded, as are the others, by a review with the sponsor. It may surprise them, for most top managers hope that all is running as it says in the QMS with some details to be adjusted. When this is clearly not so it can cause trouble: they hoped the project would be about improving, not having to reconsider fundamental principles.

There may be some immediate and obvious changes to make, and these need to be attended to or the whole project may be compromised – your earlier experience will pay dividends here too. The team will learn more than it expects about getting agreement over supposedly small and simple changes, and this will be a useful rehearsal for later wider-scale implementation. This is also a time to discover how easy it is to incorporate such small changes to the QMS. It ought to be straightforward, and so such learning will come in handy later when bigger changes are likely to be needed.

At any stage the organisation can manage lessons learned using its problem-solving methodology covered in Chapter 6 and ISO Clause 9.

*3: **Investigate sources of variation and waste*** brings together the acknowledgement that all processes vary both in themselves and through their interdependencies as they operate within the wider system. Understanding these variations provides a means for assessing their effect on the outputs for customers. The team needs to understand the impact of the cause-and-effect relationships between different processes within the system to develop theories of waste and variation at the systemic level. These theories need to be verified with data, and rigorous statistical methods are essential.

Much of this, of course, has become routine for you if you have been applying the lessons from earlier on, but is none the less powerful for that. If you have not so far become familiar with statistical process control, make sure you have experienced practitioners available to help. It's still not too late to start learning it yourself, but the complexities of most projects demand in-depth experience in statistical analysis as well.

*4. **Test theories.*** Theories for solutions to address root causes should be generated and evaluated through the Deming cycle, in a safe way that allows them to be modified or withdrawn without too much cost if the theory proves not to work effectively. Multiple tests may be required.

Testing should not lead to an all or nothing decision to accept or reject. Instead, the tests build information and knowledge about both the theory and the system within which it has been tested. It may be appropriate to test theories to disprove them, rather than just looking for positive evidence.

5. *Study results* stresses the importance of understanding the results and implications of the tests as opposed to simply reacting to them. There needs to be a period of reflection during which people can extract learning that may remain dormant unless it is developed proactively. Why was the idea that was evaluated new? It probably seems obvious now. Or, perhaps worse, why have we evaluated something we thought was new, that worked, but actually turned out to be something we used to do but has been allowed to lapse?

It's the time to relate what you have learned about the theory and practice of the QMS as it has been applied in this process – and take such learning into account later.

This is also the stage at which you can help people to draw on the systemic perspective to understand not only the direct cause-and-effect relationships found in the data but also to look for any by-products of new knowledge, which may help towards new approaches to solving other process problems. This demands that people generalise the principles of what was achieved, so that others in different processes may be able to relate to it. Such matters as communications, honest data, customer-focused measures, are all universal issues but may appear obscure if not generalised.

If the study shows the solutions were not valid, the team needs to go back to investigating sources of variation and waste once more. The better the team worked on looking for disconfirming evidence at the test stage, the more useful the theories that survive will be. Rigorous statistics are valuable at this stage too.

6. *Implement* reinforces the whole purpose of the project in terms of bringing about robust and sustainable improvements to the organisation's processes. While this may seem obvious, it is often the Achilles heel of many projects. Implementation must draw on the System of Profound Knowledge as the underlying principles for success, taking into account the wider system, the variation, operational definitions and of course the socio-emotional issues that will now be making their presence felt. The project team have the responsibility of understanding these principles in the context of the situation in which they find themselves. They then need to work by engaging local staff to introduce the changes successfully, using the Deming cycle of course. Only after a successful trial implementation should they consider wider, permanent changes to operating processes.

The relevant parts of the quality management system need to be addressed as appropriate. The presence of an auditor on the team should ensure that such work leads to a simple audit process later, as well as making the changes likely to be more enduring.

As with earlier stages you will recognise many parallels with your process management work. As also discussed before, the sequence of investigation, testing and implementation is much easier within a department than across functional boundaries. This is why it is so much better to come to a major improvement activity bolstered by such experience, rather than learning about the whole subject from scratch in the high-pressure world of big projects.

Plans and reviews of the Implement stage must be developed closely with the sponsor and maybe other members of the senior management team. It is very likely that the actions will spread beyond the original scope, and this may irritate those who have not been kept on board. Even if they support the decisions, such widening scope may have implications for their resources, and if they are in the middle of some other change it may need the senior team to act as umpire.

7. Standardise recognises that when the team has translated an improvement theory into an improvement solution, they may only have resolved the issue for the short term. To ensure repeatability over the longer term they must standardise the new process to reflect the improvements they have implemented. In some cases, this may involve a single process in a single site; in others it will involve transferring the learning across a number of sites in which parallel processes operate. Issues that are particular to each situation will arise. Contingency analysis tools can help prepare for the unexpected. The team needs to deal with these constructively so that the organisation can hold on to any new gains and ensure that it does not simply fall back into the old, suboptimal ways of working. Often there needs to be a lot of effort to develop an integrated standardisation and maintenance process, supported by clearly defined roles and responsibilities.

Work carried out in step 6 to revise the QMS will be built upon here, and can incorporate the contingency analysis studies and trials.

Whatever the team or the top management hopes for, standardisation is rarely easy or quick, and it needs to be ready to take local circumstances into account – for instance different departmental sizes, or local laws, may preclude uniformity. A 'roll-out' must not be thought of in terms of developing a solution and imposing it on every location, the later implementations must learn from the experience of the earlier ones, and indeed you may need to revisit the first few in the light of this learning.

8. Review enables the team to reflect upon the project in its wider sense. What have we learned about the organisation, and what have we learned about the improvement process? They will be able to see how the theories they held at the start of the project have developed as progress has been made. The summary report should highlight the key insights about how the organisation operates and what this might mean for subsequent improvement teams. They also need to review the process that the team went through to deliver the improvements achieved. Given that the improvement process sits within a wider support system, they must continually feed back learning about how to employ the improvement process within the recipient organisation. This process will need to evolve at least at the pace of the organisational system to remain relevant, but often it struggles to do so after the excitement of its early months. Finally, a review of what went well and what could be improved often highlights the people-centric issues that form the foundation of many of the problems that need to be overcome if the improvement programme is to yield its fullest potential.

If the project has been generated by a programme for re-registration, these reviews will form critical inputs to the wider efforts.

It is important to celebrate the completion of a project. Team members need to feel appreciated, and others will feel more inclined to participate if they see that efforts are recognised. Such celebrations may take many forms, including conferences, articles in magazines as well as internal events. Someone should be given the task of writing the story of the project, the gains and the learning, to provide input to a growing database to which all can refer.

There is more on leading major improvement projects after these ISO Clauses.

ISO Clauses

As was mentioned in the Chapter introduction there are few ISO 9001 requirements associated with rolling out standardised processes across the organisation. Nevertheless there are some specific ISO 9001 requirements for groups of processes and these are covered below in Clauses 8.1 to 8.6.

8.1 Operational planning and control (How you plan to do what you do)

You need to decide how you are going to carry out the activities that will deliver product or services and then develop those processes so they can work as you expect. In developing your processes you need to bear in mind all the other requirements of the standard that have been covered earlier (Chapter 3, Clause 4.4).

To do this properly you need to consider:

- Setting objectives for products or services (Chapter 2, Clause 6.2.1) and setting requirements.
- Whether any processes, documents, or resources are needed, particularly for producing a product or delivering a service.
- What checks of the product or service are going to be required to show it meets needs (and any particular product or service requirements for it to meet).
- What records you will need to produce to show the product or service meets requirements and that processes are working (Chapter 5, Clause 7.5.3).
- Whatever form this planning information comes in it has to be useable by your organisation.

Note 1: If you choose to put all this planning information into a document some people call it a quality plan.

Note 2: If you want to use a design process like the one in Clause 8.3 (below) you can do.

8.2 Requirements for products and services

8.2.1 Customer communication (How you keep in touch with the people who count most)

You need to develop and put in place ways for successfully communicating with your customers (Chapter 5, Clause 7.4) so that you can:

a) Provide information about your products or services.

b) Receive enquiries, orders and contracts (as well as amendments) and

c) Receive feedback from customers including any complaints.

d) Deal with any property supplied by the customer.

e) Deal with any situations when processes don't work out as planned.

8.2.2 Determination of requirements related to products and services (What do they want)

Before offering a product or service you need to ensure:

a) Product and service requirements are defined, either by you, your customer or in law.

b) You can provide the products/services to the defined requirements.

8.2.3 Review of requirements related to products and services (Can you give them what they want?)

Having captured the customer needs (in 8.2.2 above) you have to look at them and check to make sure you can supply the products and services they need. You need to do this before you commit to supply, for example before you send a tender, accept an order, contract or amendment. You need to make sure:

a) Product and service requirements are defined (see 8.2.2) including information about delivery and after sales.

b) How they are going to use the product or service (even if they don't tell you) if you know and what that means for you to supply.

c) Your own requirements – often defined by marketing departments in terms of 'look and feel'.

d) Legal requirements that apply to the product/service.

e) Any areas of the contract that are different from anything previously stated are understood.

You need to ensure that contract changes are dealt with.

If the customer doesn't put their requirements in writing then it is down to you to confirm what you believe those requirements are before you accept an order.

You need to keep a record (Chapter 5, Clause 7.5.3) of the outcome of these checks and any actions that come out of them.

8.2.4 Changes to requirements for products and services (Customers change their minds)

Where your customer changes their mind then you need to update order documents and make sure people that need to know about the changes are told.

8.3 Design and development of products and services (Introducing new products to market)

8.3.1 General

If you're not making or providing the same thing you've always done you need to design new models/versions.

8.3.2 Design and development planning (How are you going to get it to market?)

You need to plan and manage new product introduction.

Plans should include:

a) Stages of the process (including those below).

b) Design control activity (see 8.3.4 below).

c) Who is responsible for each part of the process and allowed to make decisions.

You need to make sure that the various groups involved in design talk to each other and share necessary information. You need to be clear about who is doing what in the process.

8.3.3 Design and development inputs (How you start a project)

You need to decide what information you need before you can start a design project and record this information including:

- How the product or service is going to work and performance expectations.
- Any legal requirements the product or service has to meet.
- Any information available from earlier versions to enable you to improve the design or tackle design problems.
- Anything else you need to design the product or service.

Someone needs to check the information available to make sure it is enough to proceed. You need to make sure there is no conflicting information.

8.3.4 Design and development controls (How you decide a design project is on course)

At stages you set down in the design plan (see 8.3.2 above) you must carry out a planned review of the project to:

- Decide whether the proposed design is on track to meeting stated requirements.
- Highlight any concerns and decide if any changes need to be made.

The people involved in the review must include those interested in the design stage (or stages) being reviewed. You need to keep records (Chapter 5, Clause 7.5.3) of the review and of any actions needed.

8.3.5 Design and development outputs (Drawings, specifications, etc.)

You need to produce something from each design stage that can be checked (see 8.3.4) to make sure it is in line with the information you started with (see 8.3.3 above). Someone (see

8.3.2 above) needs to approve the output before it is released to the next stage.

The output has to:

- Meet requirements stated in the input (see 8.3.3 above).
- Be suitable for people who need to use it for: purchasing (see 8.4 below) to buy materials and components; production and service provision (see 8.5 below) to make products and provide services.
- Include information about how to check the product or service is OK and is made or performing as it was intended to (or refer out to another specification that does these things).
- Identify key features of the product or service to make sure it is used safely and as intended.

8.3.6 Design and development changes (How you make sure you don't mess up a good product or service)

Where you need to change a product or service you have to capture the change need and keep a record (see Chapter 5, Clause 7.5.3). Changes must be controlled, if necessary with similar systems to those above in Clauses 8.3.1 – 8.3.5. Review of changes to designs needs to consider what impact the change will have on affected parts of the existing design and on the product already out in the field or services provided. You need to keep records (Chapter 5, Clause 7.5.3) of your control of design changes and of any actions needed.

8.4 Control of externally provided processes, products and services (Buying the goods and services you need to produce your product or deliver your service)

8.4.1 General (How you buy your stuff)

You have to make sure products and services you buy in meet your needs. You have to decide how important they are for your main ways of working and the final product or service. Depending on importance you need to decide how you will control the supplier you buy from and products and services you buy from them.

You need to choose suppliers who can supply products and services that meet your needs. You need to decide how you go about choosing suppliers, assess and reassess them and keep records of assessments and any actions from the assessment.

8.4.3 Information for external providers (How you tell suppliers what you want)

You need to tell your suppliers what you want them to provide you. Including (if needed):

- If things you need have to be approved in any way; if their procedures have to be approved; if their delivery processes need to be approved; if equipment they use has to be approved.
- If you need to approve the people to be used.
- If you have any requirements for a quality system (like ISO 9001).

You need to make sure the information is satisfactory before you send it to the supplier.

8.5 Production and service provision (The doing bits)

8.5.1 Control of production and service provision (How you manage the doing bits)

You need to plan and carry out your making of bits and the steps required to deliver a service. Things you might need to include in your plan are:

- Giving information describing important product and service features.
- Giving people any necessary instructions to make things or deliver the service.
- Making sure any equipment needed is on hand.
- Providing any equipment needed to measure product or monitor processes.
- Measuring product and monitoring processes.
- Methods to release and deliver products and any arrangements needed after delivery, e.g. service, customer care, warranty.
- How can you be sure the product is going to be OK?

You need to check to make sure that, where you can't measure or monitor process outputs to make sure they meet requirements, processes are approved as capable. This could include any processes where problems may come to light after product is used or you have carried out the service.

Approval should make sure that processes can deliver what you want. Typical things these processes need might be:

a) Ways for evaluating and approval.

b) Suitable approved equipment and use of qualified people.

c) Using defined methods and specific procedures.

d) Keeping records (Chapter 5, Clause 7.5.3) and

e) Every so often re-approving the process.

8.5.2 Identification and traceability (How do you know what is what and where it is used)

If you need to you must identify products and services throughout the process by a means you think is suitable.

Where you check products and services it must be identified to show they have been checked.

If it is important to know where components or finished products and services have gone then you must be able to manage this and record details of where these components and finished products and services were used.

8.5.3 Property belonging to customers or external providers (How you control stuff your customer gives you)

You need to take care of customer property while you have responsibility for it. You need to identify it, check it is OK to use, and then make sure you protect it until it is used or returned. If you lose or damage their property or find it is not suitable you need to let your customer know and keep records (Chapter 5, Clause 7.5.3).

Note: In this case property can include customer's information.

8.5.4 Preservation (How you look after stuff)

You need to protect products and services while you make them until you hand them over to the customer. Protection includes identification, handling, packing, storing. This includes any components that go into the final products and services.

8.5.5 Post-delivery activities (Looking after your customers)

Delivering the product/service is just part of the activity. You need to think about:

a) Legal requirements (for example product recall).
b) What could go wrong if there is a problem with your product / service.
c) How long your products and services last for.
d) What your customers expect.
e) Feedback from customers.

Note: things you need to consider are providing a warranty, whether your customer expects and has contracted you to provide maintenance, and product recycling or final disposal – for example as required in some legislation.

8.5.6 Control of changes (Take care before rushing into changes)

Before making any changes you need to review the effect of any changes on products and services to make sure they still meet requirements. You need to keep records of the results of your review of changes, who authorised the change, and any actions needed to control the change (Chapter 5, Clause 7.5.3).

8.6 Release of products and services (How do you ensure it is OK to release products and services to the customer)

Having put in place processes (see Chapter 3, Clause 4.4) and plans (see Chapter 2, Clauses 6.1, 8.1 d) and 8.5.1 c)) you need to ensure those plans are delivered. Products and services should not be delivered to customers until inspection and test plans are complete and a responsible person in the organisation has authorised delivery – unless the customer waives this requirement. You need to keep records to show what inspection and testing has been carried out and who has released product/ service for delivery.

Armed with these specific ISO 9001 requirements the roll out of process standardisation can continue across all the organisation's processes using the model described in Chapter 3.

Leading major improvement projects, part 2

We have described the methodology to be used to ensure the successful structure for major projects, and the standard provides a great deal of detail to be considered, depending upon the circumstances. Success in using the structure (the Improvement Cycle) and incorporating the relevant components of the standard depends upon leadership and facilitation of the project.

Attributes for successful improvement team leaders

Selecting the right kind of person as project leader is as important as it ever was: no amount of project management process, systemic thinking or technical skill takes away the importance of the demeanour of the person. In our experience really successful improvement team leaders exhibit a high proportion of the following attributes/skills, etc. As you scan through this list you will conclude that such people are rare, and are probably already in an important role from which it will be hard to get them released, and you will be right to do so. It's often been observed that those who are easily available are probably not right for the job. Ignoring this rule is of course one of the reasons why organisations that attempt multiple cross-functional projects usually fail to do justice to them.

Personality

- Enthusiastic, self-motivated, driven to achieve important objectives.
- Takes personal accountability/responsibility for their actions and that of their team.
- Credible in the eyes of their peers and managers, confident in their own ability and that of others to succeed.
- Able to weather tough times and knock-backs and can take personal feedback in a constructive manner.

- Challenges organisational assumptions and beliefs that are holding them back.
- Assertive but supportive and helpful to others.
- Enquiring, doesn't accept things on face value, not afraid to ask questions and listens actively.
- Seeks to understand different perspectives and points of view.
- Wants to understand why things are the way they are.
- Values learning and development for themselves and for others.

Operating style

- Demonstrates leadership: communicates a compelling vision for improvement and able to identify what needs to happen to achieve it.
- Able to follow a systematic approach to project work.
- Can organise themselves and others:
  - Can plan a work stream of activity i.e. multiple tasks and milestones for multiple people.
  - Able to delegate tasks and responsibilities.
- Can carry out basic data analysis.
- Good at communicating – both written and verbal.
- Able to present to small audiences of both peers and managers.
- Understands the organisational structure and work, knows who to ask.
- Able to lead a team of people and work as a team member.
- Seeks help as called for, both in technical subjects and personal coaching.

Relationships

- Seeks to create alignment and consensus in thinking, able to build rapport with people at all levels of the organisation.
- Engages people from the peer-to-peer perspective regardless of rank.
- Values the knowledge and experience of others.
- Empathises with others' situations, feelings, thoughts, etc.
- Able to influence others without needing executive authority.

Nobody possesses all these characteristics of course, but you will not regret using the list as a filter, and incorporating the role in your career development process.

Reviewing project progress at gateway points

The Improvement Cycle itself provides gateways at the end of each phase that are natural points for reviews. These should include representatives from the top team and senior manager stakeholders. The following questions provide the basis of an agenda for such a review meeting:

- Reminder of objectives (What are we trying to accomplish? How do we know if a change is improvement?).
- Achievements.
- Issues identified.
- Risks.
- Support required.
- Next steps proposed.
- And, of course the learning the team has gained.

Make sure that the reviews are presented on wall displays not presentations, and generate discussion among the audience, not passive acceptance.

The Improvement Cycle structure has been applied to a multitude of subjects; from improving response times in call centres to reducing energy consumption in paint spray booths, from rationalising product codes across a global chemical company to getting ships to arrive on time in port. To paraphrase a well-known cliché: 'Can it help you? Yes it can!'

Auditing

They say the devil is in the detail and specific requirements for Clauses 8.1–8.6 need to be considered by auditors. The key requirement though, remains the understanding of process standardisation covered in previous chapters. Competent auditors will start their understanding at organisational context, take it through risks and opportunities, through process standardisation, to process improvement.

Summary

In this chapter we have outlined a process for developing standard ways of working across the organisation and covered specific requirements for: dealing with customers; product and service design; purchasing and outsourcing; control of production and service provision; and releasing products and services to customers. We and our clients have found the Improvement Cycle valuable in circumstances as varied as developing global product standards and controlling the repeatability of critical production components.

Most organisations have applied or developed methodologies for significant projects, but their track records are often very patchy. The eight phases of the Improvement Cycle ensure that shared learning takes place from start to finish, and leads to reliable and enduring improvements.

5. Managing ongoing operations

Purpose of this chapter

To help you to keep the show on the road on behalf of the customers, the staff and the owners.

In designing the QMS the organisation has evaluated and redesigned its processes to include key means of monitoring and measuring (Chapter 3, Clause 4.4 c). In this chapter we look at how the organisation should capture the voice of the process and use it to standardise and improve. ISO 9001 does not prescribe how the organisation should do this but frames the overall expectation, leaving it to the organisation to decide exactly how it will do it.

Keeping the show on the road

Previously we have referred to entropy, a term not much heard in management circles, but which goes on everywhere across the universe for all time, whether we know about it or not. It is the tendency for the universe to become disordered. Mountains erode, complex life decomposes and stars explode. Or, put another way, children's bedrooms become messy, gardens are overcome with weeds and cars break down. In the same way your work processes will become less suitable for the purpose than they used to be, maybe accumulating workarounds, or perhaps training doesn't keep up with recruitment. Entropy truly is everywhere.

Life is one of the ways nature creates order out of chaos, but only as long as life lasts; it takes energy, then it decays, it becomes disordered. We are all made of elements formed in stars, put together for the moment in a marvellous but temporary natural system.

Everyone knows this intuitively, of course; gardeners, parents, decorators and drivers. Just as soon as you have decorated the hall, stairs and landing it starts to lose its newness. Kids bang the wall, the vacuum cleaner chips the door reveal, the dog leaves hair everywhere, the sun fades the carpet, someone spills coffee, and dirt arrives on shoes. It takes an immense amount of effort to keep things as they were, as exemplified by the trouble taken in ancient palaces and galleries. Even with all this effort things slowly decay, and of course the wrong sort of restoration or preservation can actually make things worse.

Because we instinctively sense this deterioration most people tolerate it in their everyday life. We do our best to stave it off but recognise that we have to keep a sense of proportion – well hopefully we do! But our customers see no need to make any allowances. They deal with our organisation without any knowledge as to whether it has been recently modernised, and expect it to deliver regardless. If the system and processes are all new it should be relatively easy to make the work work effectively (well hopefully, after any teething troubles, and if the processes were properly developed!). But the passage of time will inexorably lead to your processes becoming less effective unless energy is expended in restoring their original state. If that effect has not been maintained it may be hard to live up to your customers' expectations, particularly if competitors have been keeping up to date. You may have rework, workarounds, inspections and so on, all because the effort was not made to continually monitor and improve where it was found to be needed. Entropy is winning.

Active process management is thus the approach we take to regenerate the organisation, preventing deterioration.

It has long been hoped that a QMS would counteract this tendency to disorder. Surely if the best process is developed and documented, and management is diligent in using it, all would be well. But all has not been well. Either the documentation process wasn't well developed and was never capable – and hence is ignored – or circumstances change and it's no longer appropriate in crucial details.

Worse, when diligent managers recognised the drift and led changes to get the work adapted to better serve customers they find that bureaucracy in the MSS gets in the way.

So the future success of the revised standard depends on:

1. Developing capable processes.
2. Documenting them intelligently and applying across the board optimising interactions with others.
3. Monitoring.
4. Leading changes as needed.
5. Sharing the changes and incorporating them into the QMS.

Pay end to end attention

By now you have explored most aspects of the process, learned how it works both day-to-day and under pressure, and picked up some tools that are useful in understanding and improving it. The graphic on the next page, which we introduced in Chapter 3, acts as a reminder of the sources of variation that continually need to be managed and enables you to see what else to explore.

ISO Clauses

There is a saying: 'The proof of the pudding is in the eating' and so it is with ISO 9001 clauses. Here we look at how the organisation monitors, measures and analyses its processes to ensure they deliver what they have been designed to.

9 Performance evaluation (How are we doing)

9.1 Monitoring, measurement, analysis and evaluation (checks, analysis of data and decision-making based on data)

Developing standard operations

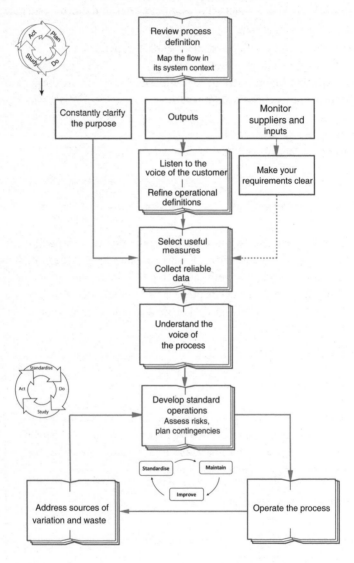

9.1.1 General

Having spent a lot of effort in developing ways of working and how to run the organisation (Chapter 2, Clause 6; Chapter 3, Clause 4.4, and; Chapter 4, Clause 8.1) you now need to put another plan together to make sure:

a) Things you make are made how you say they should be made and services are provided as you said they should be.

b) All the checks you built in to processes earlier are being carried out, and

c) Over time you get better at doing the things you do.

Your plan needs to record how you intend to do this and any number crunching involved.

9.1.2 Customer satisfaction (Are your customers happy?)

One of the things you have to do is look at information that tells you what the customer thinks of what you have provided them with and whether it gives them what they want. You need to describe how you are going to do this.

Note: There are many ways of finding out what the customer thinks including: asking them; information they give you about product and service quality; asking users what they think of the products and services; looking at reasons why customers no longer use you; if the customer tells you they're really pleased; the number of times customers demand money back and what your distributors tell you. It is also essential to get close to the actual experience of varied customers; you will learn lots about what you could be doing, even if they are relatively satisfied. Beware of putting too much trust in surveys, particularly those conducted by third parties. A lot of trees can fall in your forest (Chapter 1) and not be detected by surveys.

9.1.3 Analysis and evaluation (In a darkened room with data)

All processes generate data and you need to capture relevant data and analyse it. By analysing data you are looking at how effective systems are for delivering:

a) Products and services that meet requirements.
b) Customer satisfaction.
c) An effective and efficient quality management system.
d) Plans that work.
e) Management of risks and opportunities.
f) An effective and efficient supply chain.
g) Information needed to improve the quality management system.

9.2 Internal audit (An independent look at how well the system works)

Every so often you need to check that:

a) What you are doing is what you planned to do; that it meets the requirements of ISO 9001 and any other requirements you came up with.
b) People are following the system and doing the things that need to be done to keep the system up to date.

You need an audit plan that shows what you are going to do. In putting the plan together you need to concentrate on the important areas of the business, on stuff that could jump up and bite you if it goes wrong, oh and where you have had problems in the past look there, too.

You need to say what you're checking against, how often you are going to check, what each of the checks will cover and how you're going to go about it.

The people you choose to do the checks have to be fair to make sure the checking is unbiased – you can't mark your own homework.

You have to write this all down, including who does what in putting the plan together, doing the checking, writing reports and filing them. Those records you've just filed away, you need to take care of them (Chapter 5, Clause 7.5.3). If, when you're checking, you find something wrong you need to tell someone who has the necessary clout to fix it – sharpish. They need to make sure it doesn't happen again.

Someone needs to check that any fixes have been done and write down what they find (Chapter 6, Clause 10.2).

9.3 Management review (The top team looking at how well the system is working)

9.3.1 General (The introduction bit)

Still a responsibility of the top team. Every so often they need to look at how well the system has met your requirements (Chapter 2, Clauses 5.3 and 6.2) and whether it is still right for your needs. They need to consider any chances to improve the system and whether they need to make any changes to the system including policy (see 5.3) and objectives (see 6.2).

9.3.2 Review input (What the review should cover)

The top team needs to consider at least the following information:
a) Any actions raised at previous reviews (see 9.3.3 below).
b) Have any changes been identified at a review of context (Chapter 1, Clause 4.1, 4.2).
c) Other system measures such as:
   1. Feedback you have received from customers (see 9.1.2 above).
   2. How you're doing in meeting your objectives (Chapter 6, Clause 6.2).
   3. How well processes (Chapter 3, Clause 4.4) are working (based on process measures) and how well products meet requirements (Chapter 4, Clause 8.6).
   4. How much time and effort are you spending on dealing with quality problems and where you are with your problem-solving (Chapter 6, Clause 10.2) – how many are open, how many closed, etc.
   5. All the other controls that you have in place to check you are on track (covered in 9.1 above).
   6. Internal (and external) audit results (see 9.2 above).
   7. How your product and service suppliers are doing (Chapter 5, Clause 8.4).
   8. Where you are with your corrective actions (Chapter 6, Clause 10.2) – how many are open, how many closed etc. The same goes for actions to prevent or mitigate risks (Chapter 2, Clause 6.1, and Chapter 5, Clause 8.1).

d) Do you have enough resources (people and equipment) to deliver your processes (compared to the original plan in Chapter 2, Clause 6)?

e) Do the results indicate you have done enough to manage those risks and seize the opportunities you identified (Chapter 2, Clause 6.1)?

f) Are there any suggestions for improving the system (to feed into the process described in Chapter 6, Clause 10.3)?

9.3.3 Review output (What comes out of the process)

The top team decide what actions are needed to:

a) Improve products and services, the system and any processes (Chapter 6, Clause 10).

b) Change the QMS (although why you would change it if it isn't an improvement as in a) above, who knows?)

c) Make sure there are enough resources (Chapter 2, Clause 7). You need to keep records of these reviews (Chapter 5, Clause 7.5.3).

10 Improvement (You keep moving forward or you go backwards)

10.1 General (Another general bit)

You decide what area of the organisation needs to be improved. You should go through a rigorous process to select these areas as it is a significant investment of time and effort. The overall aim is to make the customer happier to be working with you. The sorts of things to bring in are:

a) Making products and services better. This means towards satisfying current customer needs and those that you have anticipated they may need in future.

b) Fixing actual or potential problems or minimising the effects if problems occur.

c) Making the QMS better.

> All of these improvement requirements simply capture good organisational practice and are satisfied by the processes we present through the rest of this chapter.

Use the Process Management Cycle to fight entropy

This section assumes that processes have been developed and standardised in accordance with the guidance so far.

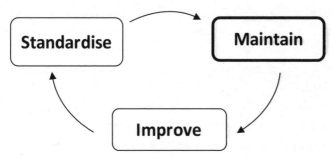

If you desire to impress your customers with your products or services, you have to ensure that work is being done as intended, and lead continual change as necessary to keep the output on target with minimum variation. Here are some key activities and processes that will help.

Use exemplary meeting management processes (this is getting monotonous...)

This is clearly a repeating theme! But, as we said before, when you consider that it is at meetings that you have the most interaction with your own and other managers it is worth the effort to set the example. Every meeting is an opportunity to use the Deming cycle to learn about how the work is working, and about how people are responding to the change efforts.

Even the shortest meeting should have an explicit purpose, a prepared agenda, use such appropriate tools as are known, be properly run, and have clear conclusions – that in turn should be reviewed at the start of the next one. Every meeting should briefly review the process of the meeting so that everyone sees the learning that emerges.

Carry out regular process reviews

If capable processes are operated according to standard operating procedures they should generate outputs that at least comply with customer needs. You therefore need to review the operations to see if standard procedures are being followed and if they are indeed producing the desired outputs. You should establish a review timetable and take advantage of the visual management systems to reassure you that all is as it should be. You should expect your managers to do the same.

Revisiting the metaphor of the forest and falling trees from Chapter 1, by doing these reviews you are increasing the chances that someone will be listening when one falls, and that learning can take place as a result.

The atmosphere of a process review needs to be one of learning and looking for evidence that, for instance, the process measures correlate to the results. It should be seen as continual Study, with the intention of identifying potential problems before they cause trouble. People should realise that their work is actually a continual experiment, with opportunities to learn from the variations as they are detected.

If you can collaborate with your internal audit people, so much the better.

A review needs to cover at least the subjects in the table opposite, which you can use as a checklist to make sure you address each of them. Other topics will no doubt emerge as you apply this list, and can be added.

How is variation being managed?

Given the generic aim of having the process 'on target with minimum variation'; use any graphs that have been developed to discuss the reality. Are they up to date, do they have live comments written on them when incidents – abnormalities and nonconformances – have occurred? This is a trigger for making sure people can tell the difference between a problem and an abnormality (or assignable cause of variation if control charts are being used as they should be). If people think that every problem is an abnormality they are almost certainly wrong, and will be making changes in reaction to all of them, tampering and making things worse. Conversely if there are no problems but they are not detecting abnormalities they are losing opportunities to learn.

Topic	Comments
Are Standard Work Instructions clear and being used?	
Is the workspace properly organised ('5 S' see below)?	
Understand operational constraints and restrictions that constrain the work.	
Look out for visual evidence of mistake proofing precautions.	
Is safety signage in place?	
Are Health and Safety requirements being observed?	

The graphic overleaf, attributed to Professor Hitoshi Kume, is adapted from the Japanese Society for Quality Control's 'Guidelines for Daily Management', a superb detailed summary of decades of learning about relentlessly enduring performance among some of their most successful companies.[1]

Most organisations find themselves in the top left corner: unstable (abnormalities present) and with some nonconformances. The whole drive for everyday management must be to get the processes into the bottom right corner: stable and conforming. You might like to talk this graphic through with your improvement facilitator: it needs to become a shared reference point.

This is one part of the process manager's job for which data are essential, along with the use of control charts (also known as process behaviour charts) to make sense of it. If there is no data, or it is being interpreted without the informed use of control charts, no one can know if a variation is abnormal (or 'special' or 'assignable', various terms are used). Thus managers will tend to intervene too often if they think every problem is special, or do nothing if they think all of the variation is common cause. We strongly advise you to become familiar with the whole subject of statistical process control as soon as possible, and you

1   Copies are available by email request: apply@jsqc.org

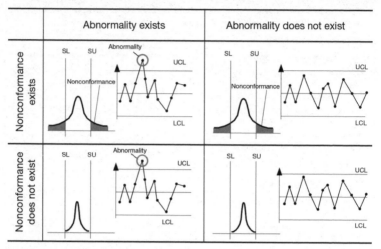

UCL: upper control limit; LCL: lower control limit; SL: lower specification; SU: upper specification

could bring in anyone in your organisation who has had the training to help you. If there is nobody available, use an external consultant, you will get your money back rapidly.

Maintain workplace organisation

Much of this subject really is common sense, in that people don't need to be convinced of the idea.

> *A place for everything, and everything in its place.*
> **Mrs Beeton, 1836–65, English cookery writer**

But it's not so common in practice. Tidiness calls for such a depth of organisational culture and persistence that it is not often maintained over an extended period. The essentials are included in the '5 S', workplace organisation discipline.

Total Productive Maintenance (TPM)

This is yet another common sense idea that's not so common. Equipment should be maintained to a standard that means it is always capable of

Sort	Straighten	Shine	Standardise	Sustain
Segregate what is needed, get rid of what is not	Designate a place for everything, and ensure everything is in its place	Clean the workplace and the equipment	Define standards for organising and maintaining the workplace	Employ systems for monitoring the level of achievement

performing to its necessary potential, so that it does not cause loss from breakdown or through gradual wear. Most of us are accustomed to the conditions of a new vehicle guarantee, which requires one to have the car serviced at defined intervals by an authorised agent. Modern vehicle systems often detect how the car is used, and issue the service instructions accordingly on when to have it serviced based upon the needs. However, such discipline is alarmingly rare in many commercial environments, where equipment is run until it fails, or buildings are neglected until actual damage occurs as a result of rain leaking in.

By applying the principles of understanding the process and its variation, TPM enables operators to predict the maintenance needs and build up a history of economic justification for appropriate changes that are improvements. PMI's video, by Dr Don Wheeler, 'A Japanese Control Chart' provides an elegant illustration of this approach.

Communicate continually and diligently

Previous editions of ISO 9001 had disproportionate emphasis on document control and records. In the 2015 edition this imbalance has been addressed. Nevertheless there are still specific requirements for control of documented information (in the standard parlance) – procedures and records to you and me – and these are covered in the following clause summaries. The first clause is about communication and is an attempt from the standards writers to capture the need for organisations to put together an internal and external communications plan.

ISO Clauses

7.4 Communication

As an organisation you need to decide about appropriate QMS communications including:

a) What you are going to communicate about?

b) When to communicate (this can be regular or as and when needed).

c) Who you are going to communicate with (the work done in Chapter 1, Clause 4.1 will help).

d) What means of communication you will use (broadcast, website, verbal etc.).

e) Who will communicate (this could be part of the requirements in Chapter 2, Clause 5.4).

7.5 Documented information (How you manage documents you need)

7.5.1 General

You need to manage any documents needed for your system. Documents can be in any shape or on any media and are divided here into two types:

1. Procedures – required to capture how things should happen as a plan.

2. Records – that capture as evidence what has happened.

In the rest of this book we refer to documents (to cover both types), procedures and records. These requirements apply to:

- Documents this Standard says you must have.
- Any other documents you need to enable you to do what you do.
- Records you need to show the system works.

Note: The amount of documentation you need will vary depending on:

a) How big your organisation is (bigger tends to mean more).

b) How complicated your organisation is. The way of working may be complex and there may be many different ways these processes work together (more complex tends to mean more procedures).

c) How capable your people are (more capable tends to mean fewer procedures).

7.5.2 Creating and updating (How you produce procedures and records)

You need to ensure that all documents produced for the system are:

a) Identified and described (so people know what they are looking for).

b) In a suitable format for their purpose (hard copy, intranet etc.).

c) Are initially and periodically checked to see they are suitable for use.

7.5.3 Control of documented information (How you ensure it is suitable)

7.5.3.1 Relevant documents must be:

a) Available where they are needed.

b) Protected from alteration (especially records). This also applies to documents that come from outside the organisation.

7.5.3.2. You must be able to bring documents (especially records) to hand, to read them and so have to control how they are:

a) Distributed, how people get hold of them and use them.

b) Stored and kept safe including reading them.

c) How you manage procedure changes (and occasionally records) (e.g. version control).

d) Kept and, when no longer required, disposed of.

All of these requirements capture the need for an organisation to manage deployment of policy, strategy and plans (covered in Chapter 1, Clause 4.1 and 4.2, Chapter 2, Clauses 5 and 6) by communicating and documenting expectations. All these activities need to be managed to effectively implement a QMS and organisations also need to consider change management in the event the original plans change.

Effective communication

As we have previously asserted, if people are to make a meaningful contribution, they need clear goals with which they agree. They also need to know what is going on, both good and bad, and they need to feel heard – that their bosses also know what's really happening at the grass-roots level. There are two interdependent ways to achieve this:

Broadcasting. This is one-way flow, and it is appropriate for information on the big picture, perhaps of company performance or market changes. In all too many cases, it is launched but allowed to lapse, perhaps because the company only spreads good news except when things are so bad that cutbacks are on the way.

The Deming cycle should be the basis of developing an effective broadcast route. What are you trying to accomplish with it? How will you know if a change is improvement? People can be asked what they would like to know and in what format. They can be asked what they think of existing information routes.

Although in principle effective broadcasting depends upon a flow from top management, local managers can take the initiative if this flow is absent. A thriving department or operation in the midst of an otherwise dysfunctional organisation is usually characterised by the local management having taken the initiative to share what they do know, even in the absence of a company policy. In our experience, this local initiative will generate requests for news about the bigger picture, which should be negotiated with top management.

Two-way communicating. The word 'communicate' is rooted in the concept of community, of dialogue, sharing and mutual help. The managers should always lead the process; if they don't do so informal leaders will fill the gap. This has often been the basis of union power.

Many well-established processes can generate two-way communication at work:

Team briefing. This is a regular, frequent, cascaded conversation between the line boss and their people. Its purpose is to ensure that staff know about the organisation and its performance, and that management at each level hears about what has been going on in the work. The top-down content is provided by news cascaded from the top management's own team brief, with appropriate additions and subtractions at each level of the hierarchy. The local information is generated at the meetings between line managers and their staff, updating each way, and ensuring mutual understanding.

The team brief should be part of the discussions at the visual display boards, integrated with other communication and information flows. In this way it reinforces the authority and credibility of local managers. It is easiest when everyone works in one location, but can be successful over the web and teleconferences if the tools are used constructively, and if the Deming cycle is used to refine them.

Suggestion schemes. The purpose of such schemes should be to provide a routine way for staff to make proposals for improvement or changes that arise from their work. Suggestion schemes have had a bad press over recent decades as a result of having been launched with a high profile only to be abandoned after a few months. The reasons for this failure should be clear to any reader who has made it to this part of the book. While staff members will almost certainly have many ideas to improve the organisation, and be more than willing to submit them, many organisations will probably not be able to use them. If processes are not stable, if analytic tools are absent, if the goals are not clear and functions do not cooperate and learn from each other, suggestions will not be successfully implemented and disillusion will set in.

Worst of all are schemes that escalate suggestions to the top. The pile of proposals in the boss's in-tray speaks volumes about the scope for improvement and the readiness of staff to contribute. But it also confirms the organisation's inability to put the two together and actually make things happen for the better.

But if the organisation has laid the foundations properly, suggestions have naturally flowed at the rate of up to one per month per employee, of which more than 95% are implemented. Although this is a bit surprising at first sight, for surely a good organisation hasn't much scope for improvement suggestions, it does actually make sense. No system can ever be perfect, and the situation is changing continually. The best people to notice this are those who see how the work works, and they are most likely to contribute if the culture is right for it. If two-way communication is effective, if local staff and managers have principles and tools to guide their thinking, many suggestions will arise and be implemented locally. If the organisation can learn across its internal and geographic boundaries, then they can be applied more widely and people get wide recognition.

Such communication is a vital aspect of an organisation becoming more self-managing, or self-organising. If, as we generally find, staff want to help, they need to understand goals and progress if they are to be able to contribute with a purpose.

Quality circles

These are workplace teams focused upon analysing their work and developing continual improvements. Originally developed in successful Japanese companies, they became very fashionable in the West in the 1980s. Indeed, both Paul and Jan led very early deployments in administrative offices in the 1980s with excellent results. When they are properly supported in a well-structured environment, they can achieve an inspiring combination of tangible improvements, skills enhancement among the participants and a constructive working atmosphere overall.

Develop skills and knowledge

You and your people need to know *how* to do things as well as what to do. You will have discovered something about the existing skill levels during the work in Chapter 1 and now need to build a robust and continuing programme to ensure everyone is able to make their most effective contribution.

The primary purpose of developing skills is to enable everyone to be effective at their work. Learning activities should be mainly focused on what people need to know, and what they can actually do in their current job. This is a wider focus than just training, with its potential overemphasis on the classroom, or on completing an online curriculum and passing the test. It needs proper preparation for the delegates, and support to them after training in trying things out.

Generalised knowledge. This is the information about the company, its products, processes and policies. All recruits, including temporary ones, need this information. In these days of outsourcing, it can be important that subcontractors who are representing the organisation, or are an integral part of its operations, also have this knowledge so that they can act in accordance with the culture of the overall system.

This kind of development should, in large companies, be the responsibility of the function of HR, and you should certainly call upon them. If it does not exist it can be developed and tested (the Deming cycle again) on existing staff. In smaller organisations you may have to do it for yourself, but you will not regret the time and energy invested because the consequence is that people are more ready and able to pull together. Health and Safety requirements may leave no option but to introduce some induction training if it is absent, but you can use the opportunity to do more than the minimum.

Remember that in developing induction training, apart from whatever the new staff experienced during recruitment, it is their first exposure to how the company works. There is only one chance to make this first impression, so it should not be the poor relation it often seems to be.

Oh, and we have often found that existing staff have been poorly inducted, so they should be the first recipients of such training. Indeed they can be effectively used in both developing and delivering such changes.

Induction training must feature the QMS as it applies to the process work, making it clear that the QMS is a live guide, not just documentation that can be ignored.

All of these practices are addressed through ISO Clause 7.1.6 on organisational knowledge – covered in Chapter 3.

Job-related skills. These are necessary to carry out the process or task, such as machine operations, and IT packages or techniques such as negotiation or interviewing. There will also be other requirements such as first-aid and health and safety. The initial process review will have exposed most of the skill needs.

Process management and improvement skills. These are concerned with understanding how the work works, and how to analyse it so that it can be improved. There are basic levels of competence that you and your people need to develop, and suggestions for the tools that will help are given at the end of each chapter. Be sure, as leader, to learn first and be seen to practise, and you thus set an example to the rest of your team.

Change demands leadership as well as management; you need to be the one to decide what is appropriate for your people, and this knowledge can only come from your practice.

Skills-related activities are covered under ISO Clause 7.2 covered in Chapter 3.

Monitoring personal performance

In large organisations managers are required to participate in some kind of performance management system, as both assessors of their own staff and in being assessed themselves. In many cases such appraisals provide the legitimate basis of training and development planning, which can be positive. In some cases assessments or appraisals are linked to pay, which can have a negative impact on general motivation. Combining appraisals for potential and development with assessment of performance for salary usually undermines both.

As with all other work (and the hint lies strongly in the name performance management *System*), it needs to be seen as a process with a purpose, and an explicit flow. It also needs to be understood in relation to its wider system. Furthermore, the Deming cycle should be applied to the process on the basis of evidence.

We are sceptical about the theory behind using such formal systems to distribute salary increases or bonuses. Our experience is that the overwhelming majority of people wish to do a good job for their employer or customer. If leaders do their job as described in this book their people will want to make the work work properly. Being able to make a difference to one's working conditions appeals to one's intrinsic motivation and is much more effective and enduring than receiving monetary rewards in response to the achievement of a target. The fact that such targets are usually in themselves arbitrary and actually not within the complete control of the person being assessed only adds to the potential disillusionment. A few individuals get some kind of reward, whereas many who have done a decent job get nothing.

What is called for is leadership from line managers. This means engaging staff in the goals of the organisation, tackling the barriers in the way, and recognising their contributions. People are individuals, with widely differing needs and wants, responses to their workplace, family environment and so on. This means that leaders have a never-ending job to try and understand the individuals and respond accordingly to each. This is of course time consuming and often difficult, but rewards all parties when done well. Setting targets, appraising people, categorising them, and rewarding by formulae do not equal leadership and does not encourage them to contribute their best.

We recognise that it can be very hard to go against the grain of an organisation's payment system. Forty years ago a manager could go for years without a proper review meeting with their boss, performance management systems were rare, and a few headline examples seemed to work for a while. Now they are everywhere. They seem to gather a vast bureaucracy of HR functions and specialist contractors, and now have a life of their own, in spite of there being little evidence that they motivate people and get them to achieve more ambitious targets.

We suggest that you attempt to understand the performance management system in your organisation by using process management concepts and tools, and engage in dialogue with their leaders to apply them in the least bad way. Some parts, such as routine skills appraisal and development planning, are needed, but using them to rank people would be best given up, just like smoking. If you work on the principle that most people would like to do their best for their employer and help them to do so in a structured way, you will often be rewarded.

Your own approach to leading

All the above methods require coaching, collaborative, behaviours by the senior managers. They involve asking questions about how things are organised, what the data tell us, and using the Deming cycle to share theories and jointly assess their validity, and to keep away from personality-based enquiries.

Our experience is that this is much the best way to get the best contributions from all staff. At heart, most wish to do a good job and to contribute to improving their place of work.

However, this is not to advocate a detached, impersonal approach. Leadership is about carrying people with you in spirit, exciting them about the value of what they are doing, not just as process operators going through the motions. Your sense of humour, empathy and optimism are just as important in a process-based world as in the hierarchy world of old. Be aware that your mood sets the mood of the department, how you hold yourself as you go through the day is important. If you look subdued and tentative for whatever reason your people will most likely interpret this as something to do with the prospects for their jobs.

You need help to develop these behaviours. It can be very valuable for you to find a third party with whom to share your doubts about your work – there are sure to be some.

Process management maturity

You can make a rough assessment of progress by revisiting the matrix introduced in Chapter 2 and comparing your comments. You have learned a lot, so don't be surprised if you change your mind about what you recorded earlier, but don't lose those comments; such a comparison is valuable when you are talking with others.

Use the matrix opposite with auditors to demonstrate your understanding both of what goes on, and what you are wishing to improve.

Auditing

In this chapter we complete the Deming cycle for individual processes. Auditors should be familiar with a range of feedback mechanisms for processes and be able to analyse output data. Auditors should feedback on the choice of method/tool and on the effectiveness of actions chosen to seize the opportunities the data present.

Organisations ask auditors to go to the *gemba* and assess how well processes are implemented, standardised and how effectively they deliver products and services that satisfy customers and other interested parties. Our audit process needs to be based on curiosity and a desire for improved process knowledge. Each potential nonconformity from audit is the start of a dialogue leading to shared understanding of whether a system improvement opportunity exists.

Level	Meaning	Comments at time of first study	Comments after establishing process management
1	The key processes are identified.		
2	Ownership of them has been established, and their purpose is understood.		
3	They are formally flowcharted/ documented and standardised operations can be seen.		
4	Appropriate and visible measures are used to monitor the processes and enable learning.		
5	Feedback from customers, suppliers and other processes is sought and used as the basis for improvement.		
6	An improvement and review mechanism is in place with targets for improvement.		
7	Processes are systematically managed for continual improvement, and learning is shared.		
8	The processes are benchmarked against best practice.		
9	The processes are regularly challenged and re-engineered if required.		
10	The processes are a role model for other organisations.		

Summary

In this chapter we have dealt with the 'Business as usual' aspect of the organisation – the aspect that most affects the everyday customer experience. However, the effectiveness of the monitoring, measuring and analysis activities and compliance with ISO 9001 clauses covered in this chapter are totally reliant on how well the system has been set up in the planning phase covered in Chapter 3 and ISO Clause 4.4. There are no short cuts.

Much of this is accepted readily by clients and their managers: they often regard it as common sense. Indeed, many senior managers get impatient and assert that they have heard it all before, and indeed that they are doing it. But in our experience it's rare to find all the practises being diligently followed across time and space throughout the organisation. And if a component is missing, or is not being integrated, the whole system is undermined – it's only as good as the weakest link.

A properly conducted audit of the actual work will rapidly expose the gaps and shortcuts, and this should be the basis for revisiting all the components, and understand what is lacking and why. It is likely to call for a revisit to Chapter 3 to rebuild the system.

6. Dealing with problems

Purpose of this chapter

To help you use a rational approach of learning and analysis even under the extreme pressure of urgent problems.

Anyone who has never made a mistake has never tried anything new.
Albert Einstein

In this chapter we deal with nonconformance arising from the ongoing operations. The requirement from ISO 9001 is to learn from mistakes and prevent repeat occurrence where possible. Nonconformities are treated as opportunities for improvement of the quality management system by incorporating lessons learned in revised processes.

Not making a drama out of a crisis

No matter how secure your processes, surprises happen and some of them will adversely impact the customer. But you can use surprises constructively. Organisations that solve problems in ways that reduce the chance of repetition become ever more predictable. They are also likely to be better at dealing with surprises when something nearly goes wrong. Possibly the best example of learning from problems is the civil airline industry, where for generations there have been both rigorous independent investigations into disasters and a no-fault culture of reporting near misses: incidents that threaten safety. Taken in combination this has ensured that mistakes are very rarely repeated. This is not the case in most environments we have encountered, where blame and threat combine to keep near misses undeclared, and accidents brushed off wherever possible. Even worse is if prosecution is likely for those unfortunate to be around when things really fail. Any system that includes the threat of prosecution is unlikely to learn much from its mistakes.

The context for 'Problem-Driven Improvement' is that of the Process Management Cycle. Here we complete the explanation with 'Improve'.

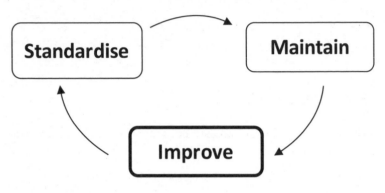

For want of a nail the shoe was lost;
For want of a shoe the horse was lost;
For want of a horse the battle was lost;
For the failure of battle the kingdom was lost –
All for the want of a horse-shoe nail.

Anon. Original is thirteenth-century German

When a crisis strikes it can be hard to keep your sense of proportion as the pressure mounts. However, you will find that the experience of what you have so far achieved will provide a foundation. When you combine the logic of the Deming cycle together with the behaviours of increasing readiness to change you can do a much better job of getting the problem resolved, and uncovering the systemic issues that lay behind it. And, as the ancient quote makes clear, there are always systemic issues, very few problems emerge from nowhere, and they are the tips of an iceberg of causal factors that have remained hidden until some unfortunate combination of circumstances came together.

The first step is to get everyone to the same state of awareness.

ISO Clauses

From the early days of ISO 9001 in 1987 it has sought to ensure that suppliers would learn lessons from nonconformity and put in place disciplined problem-solving tools to prevent problems from recurring. Over the years the emphasis in 9001 has built on this foundation to look at nonconformity prevention through effective risk assessment and planning (covered in Chapter 2, Clauses 4.1 and 6.1) and effective process design and management (covered in Chapter 3, Clause 4.4). In this chapter we address again the ISO 9001 clauses relating to the 'old' topic of identifying and managing nonconformity and applying disciplined problem solving to prevent recurrence.

8.7 Control of nonconforming outputs (Managing defects)

Where a defect is identified you need to ensure that no further work is performed on it or that it is sent to the customer. This is another part of the Standard that appears product oriented and you may think that you can skip it if you are a service function manager. But, if you consider things you 'make', from survey documents to bills to instructions and mailshots, there is plenty of scope to apply the lessons of mass production. If you seek delighted customers then none of the bills should be incorrect, the surveys confusing or the instructions illegible. Revisit the Kano Model (Chapter 1), and you will find plenty of scope for applying product-based thinking and methods in your service world.

This applies whether the defect is identified in house or after sale. Depending on what the defect is you need to do one or more of the following:
a) Correct the defect.
b) Contain the problem or recall from the supply chain.
c) Let the customer know the problem exists.
d) If the defective product/service is still usable get the customer's agreement that it can be used.

Where you choose to correct a defect you need to ensure the corrected product/service now meets requirements.

You need to keep records covering any defects identified and how you have dealt with them including who has taken any decisions.

10.2 Nonconformity and corrective action (Problem solving)

10.2.1 If something goes wrong you need to:

a) Capture the fact the problem has occurred, and
    1. Prevent the problem escalating and fix the issue.
    2. Consider the problem's impact and resolve it.
b) See if you can prevent the same or a similar problem happening again by fixing the cause(s) by
    1. Investigating the problem.
    2. Establishing the problem causes.
    3. Investigating to see if similar problems could already exist.
c) Doing what is needed to manage the problem based on the first three steps above.
d) Checking to see your actions are effective.
e) Check your original risk assessment when you put your plans together (Chapter 2, Clause 6.2), and if necessary update your assessment and your plans.
f) Decide if you need to update your QMS, and make any changes.

All of this has to be proportional to the nature and scale of the problem identified.

10.2.2 You need to keep records, as follows:

a) The problems you have identified and what you have done about them (see 10.2 above).
b) The results of your problem solving.

10.3 Continual improvement (How you get better)

You have to carry on improving your system and how it works.

Based on analysis of data (Chapter 5, Clause 9.1.3) and your top management's review of the system (Chapter 5, Clause 9.3) you need to plan how you intend to improve the QMS.

In the ISO clauses we again have no specific requirements for managing nonconformities but in the remainder of this chapter we investigate what good practice looks like.

What is the context of variation that led to the problem?

When faced with an urgent problem you will find that a context diagram such as this will help the team place the issue in the system, and guide them on deciding who should be involved with the urgent work. A key point here is to ask questions based upon process and variation, not just on the incident itself and looking for a person to blame.

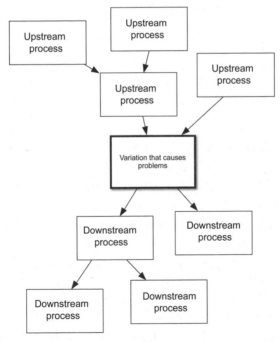

You will probably need to facilitate the group carefully to think in process and variation terms. Most people volunteer information about outputs, inputs, people, companies and departments, almost anything but process, in their first answers. The diagram forces them into process language, and language affects how people think. If they are all to understand the ups and downs of the flow of activities they have to use the language of process, expressing it using a verb–noun sequence. Sticking with this may take some persistence when the pressure is on. Some examples include 'maintain equipment' rather than maintenance, or 'select suppliers' rather than purchasing.

> *Those who never retract their opinions love themselves more than they love the truth.*
>
> Joubert, 1842

Asking about variation increases the evidence that will emerge – there will be plenty of history in the near misses but perhaps very few previous failures. A loose nail in a horse's hoof may not be a problem right now, but it indicates one that was about to happen.

Deal with immediate needs *and* get to the bottom of the process

All problems are outcomes of a process or combination of processes. As suggested above, the only way to reduce or eliminate the chance of recurrence is to understand and improve the processes. However, this is likely to take time, and time is usually short when there is a customer complaint or injured person, or a chemical leak.

Someone may therefore need to be despatched to provide immediate assistance while the enquiry is getting going, perhaps before you are sure what happened, never mind why. There's no need for them to be defensive about explaining causes at this stage however, they should just make it clear that you have a structure to get to the roots of the problem to prevent its recurrence. Get them on the road immediately after the first meeting, someone as senior as possible, especially with serious incidents. Many organisations' reputations are based upon how they respond to problems, rather than how they deliver the everyday results, as problems are an opportunity to be different in the eyes of the customer. Being really proactive provides opportunities to delight even in the toughest of circumstances.

This Problem-Driven Improvement flowchart provides the disciplined questions needed to deal with the immediate consequences and generate learning. Other useful approaches are also available, including the 8D methodology. What is important is to keep to the Deming cycle, and require the team to be prepared to work intensively to spin the cycle several times during the activity. Even if only hours are available, you must persist with the use of the Deming cycle; shortcuts risk failure at some later stage.

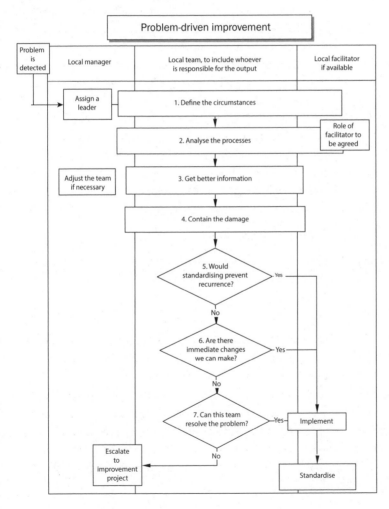

Each stage has some powerful questions:

1. Define the circumstances. What actually happened, or seemed to happen? This means asking the customer, probably visiting the part of the process in which the problem became evident, and any other area that seems immediately relevant. The term 'customer' will need to be broad in this case, as the problem may have caused an accident, or damage, or perhaps a financial loss to people who have no contractual relationship to the organisation. There is no substitute for getting to the people affected, no matter how angry and unpleasant may be the prospect of the encounter. Such anger among people experiencing failures is made much worse if they feel they are being ignored, so here is an immediate opportunity to defuse it.

Consider the communication (two-way if at all possible, see Chapter 6) you need to make, with stakeholders as well as customers. Recognise that you are probably using a different approach to dealing with the problem to the established pattern, so that will need explaining at various stages. You are seeking support for the forthcoming actions, and to influence others by your example.

2. Analyse the processes

I keep six honest serving-men (they taught me all I knew); their names are What and Why and When and How and Where and Who.

Rudyard Kipling, 1902

Use the information from 1. to analyse the process, possibly also the upstream and downstream associated processes, taking care to gather comments from operators, no matter if they don't appear directly relevant. What was varying, when did the problem happen, when did it not happen? Has it happened, or nearly happened, before? Is measurement reliability an issue, can it have happened but not been detected? Study any documentation carefully, are the standards easy to follow, are they do-able?

3. Get better information Based upon this analysis there will almost certainly be gaps in the information or data that is produced, so these will have to be obtained urgently.

4. Contain the damage. Throughout the first three steps look to identify where and how you could ensure no repetition of the incident. Look also to resolve the issue with the customer, in terms of replacement or compensation. You may find that customers tend to be less aggressive once they see that you are taking their complaint seriously, and may indeed provide extra information.

Revisit the communication you have done so far, and take steps to ensure that stakeholders you worked with in step 2 are up to date with your progress, as well as others who may have emerged.

5. Would standardising eliminate recurrence? Generate some theories (this could be in parallel with step 4.). Did the problem arise because the process was not followed? If so, test whether following the process would eliminate repetition. If this is the case, take steps to ensure process compliance in future. You may want to ask the audit people what they know. These steps may include training and information provision, not necessarily disciplinary measures. Short cuts may have been made in such things as maintenance schedules or the quality of materials being bought.

Ask 'why?' five times to get beyond the obvious symptoms.

6. Are there immediate changes we can make? Make them. If the process was being followed and the trouble happened, you have a serious case of an incapable process. Its natural variation can lead to nonconformances. The first step is, as always in this problem response context, to put a fix in place to protect the customer, in this case for the foreseeable future.

It may be possible to make process changes that will make it capable, but they need proper testing through the Deming cycle. It's thus likely that an improvement project will be needed, as described in Chapter 4. The communication you have previously done will pay dividends at this stage, as potential participants should already be aware and wanting to help. Make sure you address the QMS if it applies.

 7. Can this team resolve the problem? Is the scale of the likely project suitable for the skills, capabilities and time of the team that has been doing this investigation? If it is you can proceed to develop a project charter with them, and sponsor the improvement activity. If, as is more likely, the scope is beyond this team you will need to keep the customer protection in place while a separate project is organised.

When the problem has been resolved, use the Deming cycle with the team, and stakeholders if possible, to understand how the resolution process worked, to help run it better next time. For there is always a next time.

Note also the obsession with *process* throughout. While it is possible that an individual wilfully ignored instructions, took a shortcut, or just got distracted, these occasions are most unlikely to be the real issue. Even where it looks like an individual's 'fault' there is nearly always a systemic cause, be it in recruitment, induction, training, technology, communication, the list of possibilities is endless. Dealing with one person can keep bosses quiet, but it won't lead to better processes in future.

Over the years there have been miscarriages of justice by authorities prosecuting individuals for 'causing' problems, where it turned out that their behaviour was in fact routine, a consequence of incapability in the process. A recent example in the UK was a train guard who gave permission for the driver to start even though a (intoxicated) person was leaning against the train, and was killed. He was jailed for neglect of duty. In fact it turned out that the guard could not see clearly because of the position of the switch in the carriage. When the other guards were instructed to follow the rules to the letter the whole process was slowed so much that the timetable could not be kept to. But there was no appeal, the guard still served a prison sentence, and one can easily imagine the resentment of the others about the lack of support from their employer.

Fear will not lead to the openness that is a prerequisite for learning.

If there don't seem to be any problems, that could be a problem

No news is not necessarily good news. If no surprises are emerging over an extended time in daily operations there will be no learning. Problems can take many forms, ranging from letting the customer down to overproviding resources and hence costing too much or nonconformances that escape being noticed, in spite of best efforts in process monitoring. These are the trees that are still falling over but nobody seems to be there when it happens.

Problems are being hidden

Maybe customers are not complaining, just moving away or couldn't care. Maybe they are complaining, but issues are being dealt with and no records kept.

The task is too easy for the process

If the work works really well, in other words it is effective and adaptable, maybe it needs looking at to see how it can be made more efficient. You need to be constantly on the lookout for opportunities to match your resources with the needs of generating the output in accordance with what you know of customer needs. If you don't put pressure on to improve productivity in this rational way, sooner or later you will be pressured to do so against some kind of arbitrary target. A Quality Circle activity, as briefly described in Chapter 5, is a good way of engaging the staff in improving how their work works even when it seems to be doing well enough.

Success is the result of not following the QMS or standard procedures

People have learned shortcuts and work-arounds that are needed to get the results that are needed to get the results. All it takes is a new recruit to follow the process word for word and trouble may rapidly follow.

The process is OK now, but may not be at some future predictable time

Many factors can disrupt the routine and they don't have to be a surprise. Depending upon the organisation and its environment, workloads vary during the year. Holidays, festivals, sports events, epidemics, the weather, the list is long. You will build a picture of many of these as your first year in a job progresses, but meanwhile you should conduct a working session with your team to list the ones they have experienced, and carry out a contingency analysis on them, as described in Chapter 3. The Eight Wastes are another way to uncover opportunities.

For instance, in the travel business it's worth knowing the local holiday and sports calendars for the various destinations you deal with. There's no point providing discounted seats when there is a top match scheduled. In the food industry it's usual for the period between Christmas and New Year to be one of the busiest times, and your department's holidays may need to be rationed. On the other hand in recruitment services that time is likely to be inactive, so holidays should be required to be taken then.

You will also find much to ponder in reports and TV programmes on disasters such as oilrig blowouts, shipwrecks, plane crashes and so on. Look for the issues in process terms and you will find some sobering connections with your work.

This is moving on from problem solving to problem prevention, and once the attitudes and behaviours are established you will find it is never ending, there's no chance you will run out of possible problems to anticipate! It expands on the need for a process to be adaptable as well as effective and efficient.

People in your department will know about variation in your department, as may those in the support parts of the organisation such as HR, finance or sales. However, in our experience such knowledge is not systematically collected and shared, and each manager builds up an informal picture over the years. When they move on the newcomer has to start from scratch. A common reason we find for clients not wanting to write up a case study is because, after the presentation of the results of a project someone says, 'ah we knew that five years ago, we stopped doing X or Y after last year's budget cut'. They don't want such embarrassment to be made public.

Of course you mustn't become too obsessive and reactive! You are seeking evidence that will help to make process changes, not to get into the habit of immediate reaction to the ups and downs of everyday variation. Some days will always be better than others.

Share your learning with stakeholders

Every problem will have a natural population of stakeholders who are directly affected or just (just!) interested in it. Some of them will no doubt be closely involved with the resolution efforts outlined in the Problem Driven Improvement work, but others may only be vaguely aware of what has happened.

You should take this opportunity to keep stakeholders informed and reassured that not only has the problem been fixed, but that the team followed a discipline to ensure it does not repeat. In most organisations there are so many problems that people cannot keep track of those they are not directly involved with. They may hear about an incident being resolved, but probably wouldn't be surprised if it occurs again. If it really does not repeat they will probably have forgotten about it and won't think to credit any particular approach for the improvement.

Summary

We are reluctant to use the cliché of every problem being an opportunity, but if you are prepared to use the Deming cycle within the discipline of the Problem-Driven Improvement approach, and to *Study* its use each time a problem appears, you will find it's true. It reinforces all of the process management work you have done so far, and demonstrates that the Deming cycle is certainly better than the traditional RFA (Ready, Fire, Aim) or DDDF (Do, Do, Do, Firefight) that is so common.

So it is with ISO 9001. A fundamental principle of the use of a quality management systems standard to approve suppliers and recognise certification is that organisations have processes in place to detect nonconformity and react to it with the aim of protecting the customer and their interests. Having protected your customer the organisation should apply disciplined problem solving as described in this chapter and apply the lessons learned through the Deming cycle and feedback loops (described in Chapter 5, Clauses 9 and 10). This learning is then reassessed as part of the next Deming cycle as the organisation re-evaluates its Context (Chapter 2, Clauses 4.1, 4.2).

7. Optimising the whole system

Purpose of this chapter

To provide guidance for reliably developing a programme of change or transformation across the whole organisational system so that its performance is optimised in its wider context.

We recognise that many readers may consider it a bit cheeky to include such an ambitious concept in a book focused on the Quality Management System. It is scope creep taken much too far. But most readers will recognise the need. The old versions of ISO 9001 had worthy objectives and valuable concepts, and top-performing organisations found the process reinforced their already good practice. Many others that delegated the job of writing a QMS without changing how they worked, and did the minimum necessary to gain registration, will find that a lot tougher with the revised standard. Their transition will lead to big issues arising, and the need for a whole programme will emerge.

Escalation to systems optimisation

It is normal for apparently modest process management initiatives to uncover bigger issues that get in the way of local standardisation and the routine operation and improvement of capable processes. No process is an island, so that policies and practices implemented by others can get in the way of good management. Examples might include recruitment, training, procurement, sales, safety, and above all the imposition of arbitrary financial targets.

These discoveries are inevitable in all but the most highly developed process-managed organisations, and you will definitely know if you work in one of those!

This chapter therefore addresses the issues that, in our experience over more than 30 years, will get in the way of a smooth escalation from making a particular process work reasonably towards its outstanding performance, seamlessly integrated into the whole.

You have undergone a practical apprenticeship by what you have done so far: making your work better by understanding it and increasing readiness to change enough for modest changes to be accepted, and generating stability. You've demonstrated methods for turning problems into opportunities for permanent improvement. You may have taken this further by working at a larger scale and creating step changes in performance – and maintaining those too. This will most likely have stretched the tolerances of some of those involved and on the fringes, not so ready for change, and happy to let others take the risks. They may even be jealous! You may be well aware that further effort may generate further resistance unless the whole organisation can be mobilised.

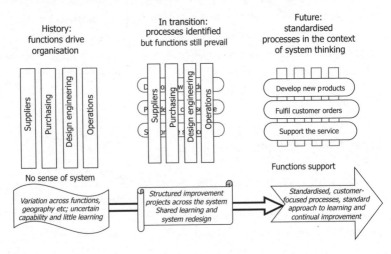

We can consider the challenges of long-term strategic transformation through the use of the concepts and models introduced earlier.

This is a subject that ISO 9001 seeks to address with Clause 4.4 b) (covered in Chapter 3) and also in Clause 10.2 (covered in Chapter 6).

Three levels of improvement

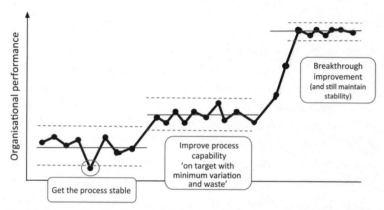

You have progressed from the left-hand side of this graph through to the right, and should feel confident that the methodologies and tools have even more potential. Many programmes throw the organisation into heroic ambitions and radical reorganisation – the right hand side – without any idea of what it needs to do just to get stable. It's hardly surprising that so many fail after a year or two. They tend to generate so many changes that the bottom part of the hamburger model – the socio-emotional issues – undermines the chance of success before the programme managers have even realised they have to take them into account. One breakthrough improvement is a reasonable expectation. Performing the trick multiple times over the whole organisation requires a lot of effort and skill – and hence leadership. That leadership needs to be learned by the senior team, it will not happen spontaneously.

Increasing readiness to change

The model for thinking about readiness to change is even more important at this scale than when you considered it within your department.

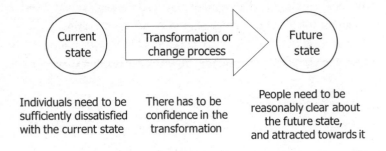

| Individuals need to be sufficiently dissatisfied with the current state | There has to be confidence in the transformation | People need to be reasonably clear about the future state, and attracted towards it |

The efforts made to involve senior people, customers, suppliers and so on in the local standardisation and improvement work will pay dividends as your ambition grows. The top management team may be ready to consider taking a new approach if they are now sufficiently dissatisfied with the performance of the organisation and they are aware of the achievements and potential of the process management approach. If you have kept good records of the situation before you started, and have accumulated evidence to show that the changes you led really are improvements, then you are on the way to addressing the future state vision – getting the organisation to be 'on target with minimum variation and waste'. That's just the start however, now we need to consider building sufficient confidence in a comprehensive transformation to be able to get commitment to launch a programme that will take years to permanently change the way work is actually done. It calls for a robust methodology to provide structure and discipline in what will always be rather uncertain circumstances that therefore require continual learning on the way. The changes to the 2015 edition of ISO 9001 put the onus on leaders to lead change and create the environment where employees are engaged in improvement.

This may be at odds with the pressures to redesign processes to enable many of them to be standardised and ready for a smooth audit. Anything that can be done to avoid the declaration of an early target date will be well worthwhile. Hence the best thing is to anticipate trouble and start to develop processes ready for recertification well ahead of the external deadline. Nothing is lost if it turns out to be easy, much is gained if the need for a big programme is uncovered. It would be nothing less than a transformation, of thinking, of methodologies and of tools, across the whole organisation.

Three-Question Model

The Three-Question Model can guide the development of a transformation strategy just as it does so well for a meeting or a problem.

While every organisation must of course create its own vision, the Four Es as we have discussed in Chapter 2 should be borne in mind.

The organisation is achieving its goals, and can demonstrate its management and improvement approach: Everywhere, Everyday, by Everyone, for Ever.

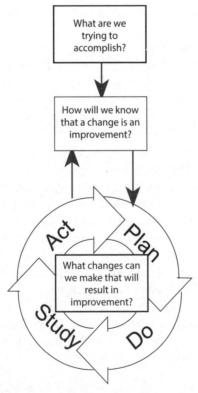

At this high level you will need both performance targets and some kind of value judgements.

The PMI Transformation Pathway described below provides structure for a multitude of changes to be tested, validated, modified and integrated as the learning and achievements progress.

True transformation must continually challenge assumptions

Any transformation is a journey that develops as the challenges are dealt with. At first it seems as though there are transactional changes that can be implemented, but it usually turns out that the barriers to doing so were hidden. Some of these may be rational, subject to evidence and testing by using the Deming cycle. But many changes will step on someone's 'emotional toes', and may falter for lack of commitment when a problem is encountered. Such crises demand careful exploration on behalf of the sponsors of the programme, and the facilitators. Some examples of these we recall include:

- Rationalising global product code descriptions, an apparently straightforward, if tedious task, necessary to present a globally consistent multi-product service on the web. However, one national operation had just been taken over and considered this demand to give up their format to be an intrusion on its independence. Much lack of cooperation led to interminable delays.
- A manufacturing company had been taken over and the new owners wanted to implement production line visual checks that included items that end users could not see. The new owners had already discovered that these check items were good indicators of overall quality, but the subsidiary managers were not prepared to allocate the resources to do this 'waste of time' work. Much aggravation resulted; the company did not survive.
- Support process managers such as finance, safety and HR who could not see themselves as providing enabling services to the manufacturing functions that actually produce the added value. Manufacturing managers resented the time taken to satisfy demands that have nothing to do with the customer and refused to cooperate when asked in the wrong way.

Such incidents did not directly lead to corporate failure, but they did indicate critical leadership issues that had to be addressed, and meanwhile they undermined the transformation programme. Time spent on anticipating such issues, and developing processes that ensure the leadership becomes quickly aware and deals with them, is time well spent.

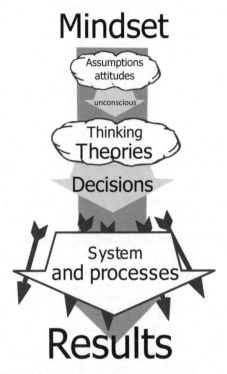

As the diagram shows, it can often be the case that people do not realise what lies behind their discomfort – their assumptions are subconscious. And even if they become aware of their assumptions, they may not be aware that they determine all kinds of decisions that should be treated more openly.

Transformation pathway

This is a step-by-step journey that, however ambitious the overall goals, builds from comparatively modest beginnings to a comprehensive programme. This initially gradual approach means that the organisation can learn about itself, deciding what kind of leaders it will need in the future, with time to develop them, both in the top team and the change agent resource, as the programme develops.

In fact, even when an organisation needs to make a commitment to a long-term programme it is important to keep a sense of proportion at first. Leaders may have decided on the need for a programme either in response to multiple critical problems or through an awareness that they need to manage for standardisation or stability. In either case, no matter how enthusiastic the sponsor of such a programme is, you can be certain that other senior people will be biding their time. They will probably have seen previous ambitious schemes fall away after a year or two. Most managers are skilled at keeping their head down until a latest fad goes away and they can get back to the old ways again.

What is therefore needed is an approach that can be self-funding, and links the initial 'act of faith' declaration to learning by doing and paying by delivering. Acts of faith are essential at first, but generally run into a crisis at some stage, so methodologies that produce tangible gains to go with improved culture have the best chance of thriving. In fact, it always takes a crisis to get people's attention to the implications of a change. In our experience only then are they prepared to take their attention away from achieving immediate goals, and engage in dialogue about what things need to be like beyond next quarter's deadlines.

Transformation development

A transformation is thus always a journey of discovery, and it can never end. At every stage there are competing factors of degree of readiness to change, knowledge about progress and challenges, and understanding of the human factors to be addressed. These combine in making each

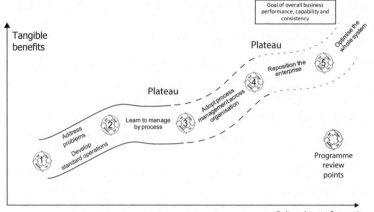

journey unique, and even when the initial goals are reached it is always clear that more can be done, and that it will need modification to the approach taken so far in order to do so.

For instance, in many transformations there is an initial programme of training and projects that produce good results and lots of enthusiasm if well done. But after six to nine months there will be turnover of staff, and new challenges. The training of replacement staff must be addressed systematically, and those who move into new jobs require new training. This may mean the transfer of training responsibility is passed from the transformation programme to the HR function, and they may not understand the implications. So unless you engage the HR managers with the spirit of the programme it will falter.

The transformation pathway addresses these issues by building awareness of the inevitability of plateaux as the programme moves on, no matter what the starting intentions were.

- For instance after a campaign of initial projects and training people become more conscious of the need to ensure that the future processes are stable and capable of maintaining the improved performance. In turn, it shows that people will find it difficult to keep individual processes stable without a supportive system, ranging from leadership behaviours to related enabling processes.

- On the other hand if the initial aim was to generate effective process management, this work will highlight urgent problems to be addressed before standardisation can effectively start.

Developing better enabling processes for particular parts of the business often generates benefits in other areas, which have yet to receive attention. It becomes clear that all of the principal processes will need to be addressed, which means the whole organisation should be thought of in system terms. Eventually nothing short of transformation of the whole system can satisfy the ambition.

The phases of the transformation pathway provide a way of thinking about progress at this wide scale that is complementary to those described earlier in relation to local process work. There are likely to be many plateaux, in different places at different times. Each is a warning that progress has stalled, and something new needs to happen to break out of the impasse. If the warning is not heeded and a revised effort is not developed things will go backwards. It is at these critical stages, as well as in launching, that external consultants can be particularly valuable.

Energy from the top is always critical. It is like the stages of an interplanetary space mission. If an organisation-wide programme is intended it must receive sufficient initial investment commitment for it to reach the stability of an orbit. Then, when the next stage is to be launched it must again be of sufficient energy and duration that people believe it will continue towards the destination. And en route it must be adjusted to keep it on track towards its goal, and with enduring commitment at the top to learn even when it is cruising. Later on in a programme it is not the lack of payback that brings the whole thing back to earth, it is not counting it, and consequent lack of attention from those at the top.

1. Launch the programme

Option 1. Address problems An improvement programme is often stimulated by failures in performance, and thus would start with projects to address critical problems (just a few, see Chapter 4), with strong support and training provided to those involved – to make sure they succeed. If this is done well, there is a surge of activities and interest with positive outcomes.

Option 2. Standardise operations Needing to make the transition to ISO 9001:2015 is just the kind of trigger for a programme of this kind. In this case the leadership decides to embark on a programme of standard operations by developing a process management approach together with its required tools. This is also generally well received by those involved, but is almost certain to expose opportunities for improvement that go beyond standardisation and have implications across functions.

Whichever the starting point, a review – Study – is called for after a few months.

2. The first plateau

After the launch, with its associated training and application work, probably some problems dealt with, there is often a pause whilst everyone catches their breath, wondering what comes next. A programme review typically reveals a series of judgements and realisations:

- If it was a problem-oriented start, there will be project benefits that are real and acknowledged, and the improvement project methodology will be well appreciated by active participants. Senior managers may well be keen to expand project activities to deal with all kinds of problems, but will not know how to select the right ones. (These ambitions may still need holding back, depending on how many top managers are ready to learn to be good sponsors.)
- If it was a process-oriented start, the methods and culture are also usually well received by those involved. However, there are probably not many tangible gains to offset against the investment, and there may be frustration building up if those problems that have surfaced are not receiving sufficient attention.
- There is more readiness to consider the needs and wants of the customer, both internal and external.
- Leaders are learning the value of the universal common language, especially in large and multinational organisations.
- There is an appetite for more training/education in the process management and improvement approach.
- There may be a new appreciation of the value of audit in helping to diagnose the need for, and nature of, changes in individual departments. A new approach is being taken to the recruitment, training and management of auditors.

- There is an acknowledgement of the value of a structured programme for the work and for good leadership of it.

However, at this stage and no matter how positive the feedback, the approach is only partially internalised; it has not become the default way of working, especially when under pressure.

People are starting to realise that stable, standardised processes are necessary to retain the gains. They are also realising that stable/standardised processes are needed if they are to reduce the need for major problem-solving projects in future.

In summary, no matter what the achievements, practice is still varied and achievements are fragile. The leaders' experience has led them to see that 'there is another way' but they are probably not seeing it clearly, and not sure what to do next.

The consequence of these characteristics is that, although the organisation has made progress, with some genuine achievements and a new confidence, it has reached a plateau. It has found out how to solve big problems, but not how to prevent them developing in the first place. Changes of policy as well as activities are needed if the gains are to be retained and built upon.

Depending upon the readiness to change of the whole organisation, the programme leaders may decide to revisit their starting point, filling in the gaps exposed no matter what was the first emphasis. This builds more evidence to support the next push forward.

3. Secure the gains, adopt process management across the organisation

The whole organisation evidently needs a comprehensive and rigorous approach to managing processes to build upon the achievements so far and be ready for more. It needs to learn about the nitty gritty of everyday process management that complements the methods and tools proven in the project work. It is likely that a rethink about the scale of the QMS is needed, particularly if it has been a poor relation so far. This is not to say that the work already started should not carry on as before – far from it. There will still be more opportunities for breakthrough improvement than the organisation can handle and if they are properly selected, supported and learned from, the properly completed ones reinforce the whole message.

The process management phase of the programme consists of policy decisions about process ownership, the QMS, extensive training in process management alongside continuing training for those involved with projects, and increasing coaching of leaders to incorporate the new approach in their everyday jobs. It is likely to ask awkward questions about functional responsibilities, management hierarchies and so on.

Many of the achievements of this phase will take the form of deepening those already in progress. These are **emphasised**:

- Project benefits are acknowledged as real and acknowledged, the method is appreciated, and projects have been repeated and **some are written up.**
- There is more emphasis on the customer, and **customers are noticing the difference**.
- Leaders realise that stable, standardised processes are necessary to retain the gains, and have **personally led such processes through crisis** and difficulty.
- Training and education includes much more content for **everyday process management**. All training in other disciplines incorporates aspects of process understanding and improvement. For example, this includes subjects such as health and safety, sales, design, projects, recruitment and finance.
- **Managers have substantially justified to themselves**, and can explain, the claim that stable/standardised processes reduce the need for major problem solving projects.
- There is **acknowledgement of the value** of a programme for the work and for good leadership of it, but there may still be frustration at the pace and many needs for support for individuals.
- The approach is internalised, and is becoming the default approach, **especially when under pressure.**
- The QMS is thought of as a useful resource by an increasing number of people.
- Much **management practice is standardised,** achievements are robust; effectiveness is seen as the route to efficiency, and capability is being translated into ability.

In summary, by this time the leaders can describe their personal role in the change, and they know there is even more potential. They are confident about what to do next

4 and 5. Reposition the enterprise, optimise the whole system

The experience of earlier phases of the Pathway will lead to an increased realisation of the limitations of the organisation's structures and strategies. Leaders and practitioners will find that continuing progress is increasingly being impeded by contradictions across the organisation. Contradictions can include managing by departmental budget, manipulation of data, and pressure on projects to deliver before they have run their trials. Any one of the traditional top-down management behaviours can get in the way. If top management does not address the contradictions, frustration builds and achievements plateau once again; they may even go downhill.

It is at this stage that the foundations you have established will bear fruit for the organisation. The common language helps people see that the principles are valid, and that the methods will help address the deepest of strategic problems. Top management teams that call upon the expertise find that they can make unprecedented changes with the help of their people and not in conflict with them as has often been the case before.

This concept of programme development enables you to be aware of the stage of your own thinking and to anticipate the greater scope and ambition that each review will reveal.

Assessing progress

Part of the dialogue with those who are reluctant to change involves the evidence of the benefits, which the traditionalists will probably downplay if they can. Throughout all this work there is a critical need to keep track of achievements and learning. Everyone, from the most junior clerk to the most senior director, needs encouragement to keep on with changing by appreciating what is being gained, both tangibly and intangibly.

The revised ISO 9001 can provide a foundation of this reference, and using it in this way will achieve multiple benefits. Excellent QMS have long been instrumental in leading organisations that have chosen to take the subject seriously and we hope that many more will find this out for themselves by taking the approach we advocate in this book.

There are also plenty of corporate assessment processes, such as the EFQM Excellence Model in Europe, the Malcolm Baldrige National

Quality Award in the US, and the Deming Prize (now the Deming Grand Prize) across Asia. Any organisation that takes the new version of ISO 9001 seriously will be well placed for any of these awards. They have all been useful, but have also been used in some cases for the glory of the prize rather than the real transformation of the performance of the organisation on behalf of the customer.

We think it most useful for an organisation to understand its own achievements first, and apply for an award when it is confident it is doing it for the right reason. You can use some of the approaches already described to judge progress and needs for change to make better progress.

How are you doing in relation to the Four Es?

The organisation is achieving its goals, and can demonstrate its management and improvement approach Everywhere, Everyday, by Everyone, for Ever.

Two components of transformation are tracked in this way, results and method. If the organisation is achieving its goals that's great but it may just be luck, it may suffer when times get hard. But if the organisation does understand the contribution of the approach we have been describing you can be very confident in predicting that the improvements will pay dividends though bad times as well as good.

We find that these four fields of judgement make sense to ordinary managers, of whatever seniority.

Review process maturity

In Chapter 2 we introduced the idea of a graduated series of levels of maturity, and hopefully you made some notes then. Now is the time to refer back, comparing how things seem to be now with how they looked then. You may find, as many others have, that you now think you were too generous at first. As you learn more about process management you also learn that the potential is more than you could ever have conceived. This goes with being a very hard taskmaster when you visit other work sites – or shops, airports, banks, sports clubs... One also has to learn to keep a sense of proportion and humour, or get a reputation for obsessiveness!

Performance characteristic visible	Contribution of major projects	Contribution of process management
Everywhere. Across the whole organisation, including strategy development, everyday work and projects.	Projects are multi-function, with strategic goals, solutions are implemented into everyday processes.	All can see how they contribute to the optimisation of the organisation, either in delivering added value to the customer/end user, or in enabling the delivery processes.
Everyday. Leaders understand and can explain the relationship between how they approach their work and the overall improved results they have achieved.	In order for the changes to be permanent, leaders have had to make links between methods behind the achievements of the projects and their implementation into daily management.	Daily process management delivers products and services predictably, responds to problems as they arise before they impact on customers.
By **Everyone.** The approach is used in depth where appropriate, and can be explained by line managers and staff routinely, not just the improvement personnel.	Active sponsorship by top management ensures they understand that the approach is different. Continual review of the improvement process means that all involved have helped adapt the approach and don't have to rely on specialists.	People are trained in methods to analyse their processes and their data, and have the will and the permission to improve locally, and to contribute cross-functionally.
For **Ever.** It has clearly been applied and developed over many years.	As the large-scale projects take root, and the approach is used again and again, participation becomes part of everyone's career development.	'This is how we do things here.' Newcomers are properly inducted, non-adopters have gone.

Level	Meaning	Comments at time of first study	Comments after establishing process management	Comments after transformation work
1	The key processes are identified.			
2	Ownership of them has been established, and their purpose is understood.			
3	They are formally flowcharted/ documented and standardised operations can be seen.			
4	Appropriate and visible measures are used to monitor the processes and enable learning.			
5	Feedback from customers, suppliers and other processes is sought and used as the basis for improvement.			
6	An improvement and review mechanism is in place with targets for improvement.			

Level	Meaning	Comments at time of first study	Comments after establishing process management	Comments after transformation work
7	Processes are systematically managed for continual improvement, and learning is shared.			
8	The processes are benchmarked against best practice.			
9	The processes are regularly challenged and re-engineered if required.			
10	The processes are a role model for other organisations.			

Four-Student Model

We introduced this concept in Chapter 1. It takes us beyond the raw impressions created by achievement or failure, by asking both 'how did we do?' and 'did we go about it as we intended?'. This pair of questions lie at the heart of all good coaching, by getting the individual to think through the 'hows' as well as the 'how muches'. A tennis player will try and recall if they threw the ball up in just the right way for the serve when the ball misses the service line. A golf player might try and recall if they kept their head down when a shot flies over the green. A meeting manager should think about the steps they took in preparation before the meeting that ended in confusion – had they the right attendees, was the agenda clear, did they check on 'what are we trying to accomplish?' at the start of the meeting, not halfway through after some confusion?

The four-student model provides a reminder that we learn when things don't go as expected, but we learn little if they run smoothly, unless we take the steps to articulate what was different to our previous behaviour.

Transformation demands three roles of management

We have seen how the basic requirement of a line manager is to ensure that their work produces the outputs required by the customer, alongside the associated demands of managing people, reporting results and so on. Many people spend much of their career fulfilling this role, responding to problems, and hopefully cooperating with major changes as they are introduced.

We have also seen how such a manager has the opportunity to escalate problems or opportunities beyond their immediate responsibilities, and participate in or lead longer-term, cross-functional projects that lead to process redesign. If they have explored and applied the process management approach in their everyday work they will find it powerful in the project environment.

Developing a longer-term strategy uses exactly the same principles, but with additional methodologies to guide the use of the tools.

The diagram overleaf is a key piece of evidence for our assertion that the System of Profound Knowledge, with constant use of the Deming cycle, can be a universal foundation for all your work. Those who have conquered the challenges of getting processes 'on target with minimum variation and waste', and keeping them there, have completed an apprenticeship that puts them in a strong position when a top management opportunity comes along.

Three roles of leaders

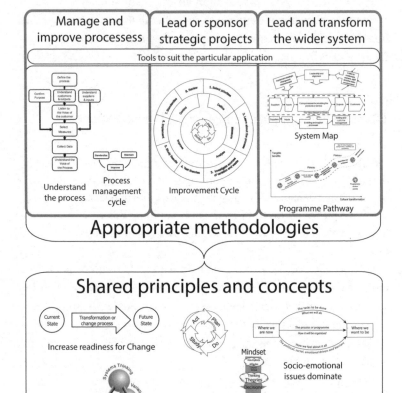

Appropriate methodologies

Shared principles and concepts

In conclusion

Introduction of the 2015 edition of the world's most widely used standard for quality management, ISO 9001, presents a great opportunity for the one million plus certified organisations to look again at some really important issues:

- A requirement to consider the organisation as a system, including its relationships and interactions with its context, for example customers, suppliers and the wider environment. There is a continuing emphasis on managing internal processes both individually and as part of the organisation's system.

- Requirements to proactively consider risk at a system, process, as well as product level, and to plan for changes to take advantage of the opportunities available, not just to mitigate the problems as they may present themselves.
- Requirements to plan for continuing improvement and transformation, not just problem-solving.

In our experience, and we have worked with impressive and successful organisations, it takes a great leader to pursue the relentless implementation of all the elements described in this book. We believe that ISO 9001:2015 may well be the catalyst that drives more leaders to pick up this baton and run with it. When every ingredient makes its own contribution properly, in combination, we have seen spectacular gains. Our colleagues and clients have experienced many epiphanies, and once that happens they are committed for life.

We have touched on some of the reasons where organisations have only a limited scale of long-term application of these methods, short-term financial or other targets being the chief culprit. But beyond this and other logical barriers is the reality that, once someone has obtained a senior position, leading in a particular way, they generally feel safer to carry on with whatever approach made them successful, rather than what they may see as taking risks with new ways in which they are not practised.

Let us reassure you, leading change using the Transformation Pathway, and integrating it with a new QMS that would sail through an ISO 9001:2015 assessment, is safe.

So we look forward to working with those who do want to take the chance, helping them commit to the approach, using the methodologies we have described to guide the tools, in the context of the Deming cycle.

As we asserted earlier, the principles are sound, the methods logical and the tools are honest.

The secret of change is to focus all of your energy, not on fighting the old, but on building the new.

Socrates

Appendix: PDSA? Deming cycle? PDCA?

This simple diagram features in one of only two graphics in the Standard, so is evidently important. And indeed, the ideas behind it are truly profound, for it represents the scientific method, adapted for practical decision-making, and is key to the success of every aspect of this book from the widest strategy to the most immediate detail.

It is a system for learning: interdependent components that must be integrated in order to achieve the goal of decisions and plans that will delight customers and so on. It depends on feedback from observation to initial decision-making, through planning what to do, implementing it and thus back to observation once more. It needs to allow for both original ideas and creativity and for the most routine of standard operations. You will find it referred to throughout the text.

With all the complexity of these concepts – it forms the bulk of those associated with Theory of Knowledge in the System of Profound Knowledge – the four short words each carry a heavy load, and they sometimes don't do it very well. This has led to problems as words can easily pick up meanings that were not intended.

The most obvious example of this problem is the word 'Check', used in the first version of the cycle developed by Shewhart and Deming, and some of the pioneering Japanese in this field, in the 1950s. This sector of the cycle concerns observation, exploration, testing, researching, and an awareness of degrees of accuracy, repeatability, bias and so on, whether the subject is an experiment, today's output or this year's strategic planning.

The Japanese founders adopted the whole PDCA concept with diligence and applied it across the board; you will find it very widely quoted in their literature. For them PDCA is almost a word in itself, conveying the combination of components rather than an acronym for the individual words.

However, when the quality management concepts were being adopted in the English-speaking world from the 1980s many people read the individual words and created their own concepts. 'Check' came to mean, in all too many cases 'is it OK or not?' It was treated as a quick step in decision-making, and often losing the opportunity to learn from the data. 'Check' suits those who are merely trying to conform to specifications, and is not very useful if you are learning how to reduce variation.

As a result of his frustration with this, Dr Deming adjusted the cycle in the late 1980s so that 'Check' became 'Study', and those under his influence, the quality management movement in the Anglophone world adopted it. It was not quick or easy: not the least of the difficulties for our company was the magnitude of the editorial changes across our course books and overheads.

However, PDCA lives on in the bulk of general literature, and its deeper meaning is often not expanded upon, so its limited implications are widespread today.

A further problem is that in many languages the first letters of their equivalent words do not begin with PDCA or PDSA, whatever meanings are given them. These examples, from Poland and Germany, illustrate that they would each need a different acronym. Not very standardised! It is confusing when some in an organisation (often senior managers) speak English, but others do not.

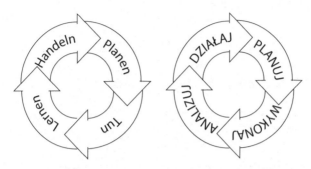

Then there are the many languages not based upon Roman scripts. In a recent conference in Tokyo we asked some very knowledgeable Asian friends to draw the cycle in their first language. We expected this would not take too long, after all we are only looking at simple words being translated. But in several cases it took two days to achieve, with much humour, as colleagues who work in English for their professional life struggle to translate 'Act' with a different meaning to 'Do'. Here are two examples:

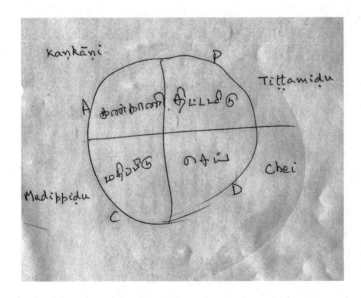

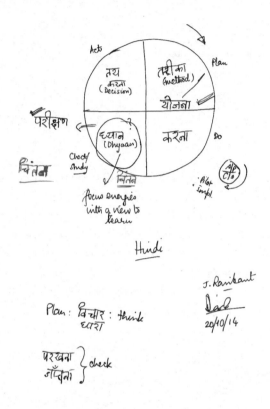

This creates non-standard interpretations of a core concept of the Standard. Of the more than a million registered organisations, many managers do not have English as their first language. Those of us who think and work in English cannot know what the instinctive meaning will be for others faced with Plan-Do-Study/Check-Act. For all of our efforts to translate we cannot appreciate their background to the words. For instance we understand from Japanese friends that 'Study' has negative associations with intensive school and university work.

Time for an international standardisation on the Deming cycle

Science has long got around this difficulty by using the name of the person most associated with a theory or concept. Ohm's Law, watts, Mach numbers all respect their originators. The Kano diagram, Taguchi's loss function and the Ishikawa diagram are all widely recognised in the same way in the quality management field. Beginners have to understand what lies behind the names, which have no descriptive meaning that suggest simple ideas. When they have learned the concepts step by step, by exploration and discussion, they are able to use the shortcut name with confidence that others will have a similar interpretation.

Many people already use the Deming cycle for PDSA/PDCA (as any web search demonstrates). We propose that its general adoption would help everyone around the world. Teachers and leaders would summarise it as a learning cycle which links observation and research to decision-making, through planning of activities and for contingencies, to execution, observation and interpretation of events, processes and outcomes, and back to decisions once again. People could then rely on a degree of shared understanding, uninfluenced by overly simple words, with those who speak no English – a significant proportion of managers across the more than a million registered organisations.

Throughout our book we use the Deming cycle unless discussing the four stages of PDSA. We look forward to feedback from both newcomers to the ideas and those who remember the origins decades ago.

Questions for discussion

These questions are designed to help stimulate thinking either for you or for you to use with your colleagues. We've grouped them by chapter and hope that you can use them to relate to your current work and what you are looking to accomplish.

The questions are structured under the components of the System of Profound Knowledge so that you can appreciate the value in thinking about all aspects of your work from each: it is often the case that problems have arisen through losing the insight that each part has to offer.

However, the real value arises not so much from the disciplines of the main subject, which can often be reasonably obvious, as from the relationships with other parts of the model. For instance a process flow chart is clearly a methodology that helps you understand the **systemic** flows, but the success of the flow is dependent upon the **variation** of its upstream processes. Its performance should by reviewed and improved by the use of the Deming cycle (PDSA) and it is essential that those people **(Psychology)** who operate it understand its purpose etc etc.

We have selected only a few, illustrative, questions. You will find many more to consider.

Chapter 1 Where are you now?

SoPK category	Observations	Insight from other parts
Systems thinking		
What seems to be the purpose of your system?		
What is the scope of the Quality Management System?		
Who are your interested parties and what are their needs?		
Theories of variation		
Where does the variation come from that impacts your processes?		
How much understanding is there of the goal of 'on target with minimum variation'?		
Can people tell the difference between a problem and an abnormality?		
Theory of knowledge		
How does your process relate to the Four-Student Model?		
Are your goals operationally defined?		
Psychology		
How might the Kano model help you understand your customers' views?		
What do you now notice about the results from recent change initiatives in terms of the 'readiness to change' model?		

Chapter 2 Summarise your system, decide on priorities

SoPK category	Observations and insights	Insights from other parts
Systems thinking		
What were the key lessons from preparing your system map?		
How well did you rate on process maturity?		
What immediate issues arose from developing your local process flow charts?		
Variation		
What have you learned about the reliability and usefulness of your current measures?		
How does the management of your QMS relate to the variation in everyday outputs?		
Theory of knowledge		
How useful and robust is the learning that emerges from current training processes?		
How easy was it to bring together the information gathered and use to **Act**: to agree on the process to be piloted?		
Psychology		
How well does current leadership behaviour relate to Clause 5.1?		

SoPK category	Observations and insights	Insights from other parts
Considering the recommended meeting processes, how do they relate to your experience and practice?		
Who has been responsible for the QMS in the past and how successful have they been at getting senior leadership support for the system/buy-in for the system?		

Chapter 3 A Pilot process: learn how to standardise

SoPK category	Observations and insights	Insights from other parts
Systems thinking		
What processes are used for planning to reduce and mitigate risks?		
How can you ensure that the Process Management Cycle becomes the default point of reference in process reviews?		
What's involved with putting the QMS at the heart of how you run your operations?		
Variation		
How can you make everyday process variation visible to all in the workplace?		
Do people realise that standardised operations are a requirement for minimum variation?		
Theory of knowledge		
How easy was it to keep learning in mind – as opposed to judgement – in your pilot?		
Can your progress meetings be better structured to enable people to learn rather than feel judged?		
Psychology		
What has the pilot taught you about increasing readiness to change?		

SoPK category	Observations and insights	Insights from other parts
What did you learn about facilitation and team make-up in the meetings?		
How did you deal with the 'bottom part of the hamburger'?		

Chapter 4 Develop step change and standardise

SoPK category	Observations and insights	Insights from other parts
Systems thinking		
How does the structure of the Improvement Cycle compare to the approach in your organisation?		
Which of the phases of the Improvement Cycle have particular influence on operational planning?		
Variation		
What did you learn as a result of seeking for 'sources of variation' rather than 'causes of effects'?		
Theory of knowledge		
What are the implications of always looking for learning by everyone before making changes?		
Psychology		
How will you develop the right kind of people available for high-pressure project facilitation?		
What further learning do you have in relation to the hamburger model?		

Chapter 5 Manage ongoing operations

SoPK category	Observations and insights	Insights from other parts
Systems thinking		
How well did you perform in relation to the 5S analysis?		
What kind of changes have you needed to make in your documentation?		
Is everyone now relating well to the Process Management Cycle?		
Variation		
How are people now making the distinction between problems and abnormalities?		
Theory of knowledge		
Are people regarding everyday variation as opportunities for learning?		
How well do your training processes lead to useful learning?		
Psychology		
How is the learning about processes in general influencing the many aspects of communication?		
What did you learn about facilitation and team make-up in the meetings?		

Chapter 6 Dealing with problems

SoPK category	Observations and insights	Insights from other parts
Systems thinking		
How easy is it to ensure you can look at the whole context when dealing with problems?		
What are the experiences of recognising and controlling any faulty outputs?		
How easy have you found it to insist on the disciplines of the Problem-Driven Improvement process (or similar)?		
Variation		
Are your problems typically the results of abnormalities or of common causes – and are you sure everyone knows the difference?		
Theory of knowledge		
Consider how often it is that the investigation results in learning for improvement.		
How often has a problem turned out to be an extreme example of something that had already happened but was not learned from?		
Psychology		
What has been the experience of keeping everyone well informed about progress and the origins of the problem?		

Chapter 7 Optimising the whole system

More assertions than questions, but good for discussion nonetheless

SoPK category	Observations and insights	Insights from other parts
Systems thinking		
The scope always grows.		
The idea of optimising the whole system (as opposed to adding together the individual parts) is often hard to deal with.		
The Transformation Pathway is a system for change.		
Variation		
The goal everywhere should be 'on target with minimum variation', not just 'conform with requirements'.		
Theory of knowledge		
Honesty in measurement, recording and interpretation is essential but hard.		
Review process maturity for insight not pleasure!		
Psychology		
Always seek to expose assumptions.		
Every activity has requirements for careful attention to 'increasing readiness to change'.		

About the authors

Jan Gillett first met Susannah Clarke several years ago when she dealt with the consequences of some problems involving her company's e-Learning programmes being customised for a leading PMI client. Her positive approach and unflinching honesty, together with her success in resolving the highly charged issue (which had plenty of scope for big losses), set the tone for the years that have followed. She later joined PMI as a Non-Executive Director, and in 2013 became a full time Managing Partner.

In order to develop the strategy to integrate the revised ISO 9001 standard into PMI's practice, in true PDSA fashion, Jan and Susannah began with study. They wished to understand more about the changes: how auditors, quality managers and organisations felt about them; where their potential areas of pain were; whether it was evolution or revolution; how PMI's principles and methods could add value; and if this really was an opportunity to achieve the ISO Committee's true intent, to provide organisations with a standard that will 'help to improve its overall performance and provide a sound basis for sustainable development initiatives'[1]?

Through 2014, a wide set of data was gathered from customers, colleagues, associates in Asia – particularly Professor Yoshinori Iizuka – and PMI's partners the CQI/IRCA. This trail led to Jan and Susannah meeting Paul Simpson early in 2015.

It turned out that Paul was just as keen to bring the standard to the improvement and transformation movement as Susannah and Jan were, and together they have enjoyed discovering many links of principles

---

1   From *ISO 9001:2015*, Introduction, 0.1 General

and methodologies. This book is their first collaboration but they look forward to many more!

Susannah Clarke

Susannah Clarke is a Managing Partner at Process Management International (PMI) and a specialist in the field of executive and performance coaching. Susannah has worked extensively in the learning and development sector, starting her career with NatWest Markets in the City before spending seventeen years with GSK as a consultant. From 1989 to 2007 she established and led a consulting, learning and development company, specialising in training outsourcing, programme and project management, learning management system implementation and eLearning development. In 2007, following the successful completion of the sale of her company to a US competitor, she developed a portfolio career as an independent consultant. In addition to working on opportunities for major training outsourcing projects she completed her ICF/EMCC accredited executive and performance coaching programme and delivered coaching programmes with a variety of organisations from financial services to The National Trust.

In 2011 Susannah joined Oracle University as Partner Director for EMEA. She had responsibility for the development and deployment of the EMEA partner strategy, leading a team of in-country partner managers, quality of partner delivery and commercial results. After 5 years of working with PMI as a Non-Executive Director, in 2013 she joined the business as a full time Managing Partner.

As co-author of *Implementing ISO9001:2015* she brings more than 30 years of experience leading, managing and consulting across different organisations. Susannah is a Non-Executive Director for RPC UK (a Project Controls Specialist and Oracle Primavera P6 EPPM Specialised Gold Partner); she regularly addresses conferences, writes blogs and publishes articles in leading process and quality focused publications.

Paul Simpson

With a first degree in mechanical engineering, Paul Simpson started his career in quality management with Pirelli. Throughout his career he has applied and developed quality knowledge and skills in a range of industries and organisations and has extended his postgraduate qualifications into the fields of marketing and business as well as professional areas of auditing, health and safety, and quality. Paul contributes to the quality and risk professions as a Fellow of the Chartered Quality Institute/Chartered Quality Professional and as Specialist Fellow of the International Institute of Risk and Safety Management (IIRSM) and IIRSM council member and has provided articles and training and delivered events promoting quality, risk and management.

Paul participates in UK and international standards development and committees including as a technical expert for ISO Standards Technical Committee TC 176 (quality management and quality assurance); BSI's QS/001(quality management and quality assurance procedures); HS/001 (occupational health and safety management); AUS/001 (revision of ISO 19011) and CAS/001 (conformity assessment).

Jan Gillett

Jan Gillett joined the Pilkington Group 1968,working in sales, marketing and customer services, in 1983 becoming Managing Director of subsidiary Kitsons Insulation Products Limited. In 1986 he become Managing Director of Sketchley Textile Services and it was in this role that he responded to the invitation of their major customer Ford motor to take up Deming's approach to Total Quality Management. He led the consequent company-wide transformation across ten locations and 2,000 people, meeting Dr Deming several times during this work. He also sponsored a division to achieve the BS5750 Quality Standard.

In 1990 he established PMI in the UK and, together with Jane Seddon from the mid-1990s, bought the UK PMI business from the Americans and led its development in working with clients in every sector, and in every continent. He became a Director of the British Deming Association, and also Chair of the Alliance of Deming Consultants. Attending several of Dr Deming's famous four-day seminars, he provided contingency cover for Deming's last event in Europe, in Zurich in 1993.

Jan Gillett is a Fellow of the Chartered Management Institute and a Member of the Chartered Quality Institute. He has also been a non-executive director in several organisations. He has presented at many conferences across Europe, Asia and the United States, and is a visiting lecturer to the University of Warwick's Warwick Manufacturing Group.

About PMI

Process Management International

PMI is unusual in the world of business improvement consultancy and training in basing our practice on explicit theoretical foundations – in our case Dr Deming's System of Profound Knowledge. The company was founded in 1984 and has continually developed programme, consulting and training services to put theories into operation for the benefit of clients wishing to achieve better results. In that time, the business has helped transform the performance of many organisations and served thousands of people through its in-house, public and online training services.

We are proud of our relationships, many of which date back to our foundation. People tell us that our special appeal is our combination of logical process analysis with key aspects of interpersonal and cultural behavioural change. This helps clients to address their organisational challenges, thereby enabling them to make effectiveness and efficiency gains that have collectively amounted to several billion US dollars. But in addition to that, and the reason why people remember their encounters with us for so long, our approach is rewarding personally and emotionally, enabling them to make sense of their work.

Today, we are headquartered in the UK and we work around the world with clients in a wide range of industries. Our practice has diversified over the years but the founding principles of people-centric process improvement and decision making using data are still applied no matter what the setting.

As partners of the UK's Chartered Quality Institute, the oldest such body in the world, PMI offers services and support that achieve rapid, effective and permanent results.

www.pmi.co.uk
Villiers Court, Birmingham Road,
Meriden Business Park, Meriden, CV5 9RN
Tel: +44 (0)1676 522766
Email: info@pmi.co.uk

Further reading

Many shelves of books line the business section of any large bookstore. We have found the following stimulating and useful.

Making Your Work Work complements *Working with the Grain*, Gillett J., Seddon J. (2012, second edition), Process Management International 978095460565. We describe the principles and challenges of organisational transformation, and explore programmes to help make a success of it. In particular, we explore parallels between our approach and the field of Natural Systems thinking.

About Dr Deming

W Edwards Deming Institute: www.deming.org is the official site for the man, his history and current activities about his approach.

Out of the Crisis, Deming, W. E. (2000, second edition), MIT Press. 9780262541152
Deming's definitive work from the mid 1980s. Open it at any page and you will find valuable insights. Although it is inevitably inclined towards the US in the 1980s, it remains highly relevant.

The New Economics, Deming, W. E. (2000, second edition), MIT Press. 9780262541169
Originally published in the year of his death, this book is an easier read and contains his explanation of the System of Profound Knowledge.

The Deming Management Method, Walton, M. (1992), Mercury Business Books. 9781852521417
An accessible guide to putting Deming's ideas into practice in terms of 1980s culture.

The Deming Dimension, Neave, H. (1989), SPC Press. 9780945320364
Another narrative of thinking and applications of Dr Deming's concepts, from the point of view of a British author.

Systems thinking

Profit Beyond Measure, Johnson, H. T. and Broms, A. (2000), Prentice Hall. 9780684836676
Tom's personal exploration of the implications for top managers in accepting the lessons from Toyota, and transforming towards more emphasis on how things are done as opposed to only thinking about targets and results.

Fifth Discipline, The, Senge, P. M. (2006, second edition), Random House Business Books. 9781905211203
Peter Senge's classic work from the early 1990s updated and still just as relevant. More insight into how systems work. The companion *Field Guide* is packed with useful methods.

The Toyota Way, Liker, J. (2004), McGraw Hill. 9780071392310
One of several books from Jeffrey Liker outlining Toyota's practices for anyone to use. Others include analysis of technical development and of training.

Toyota Kata, Rother, M. (2009), McGraw-Hill. 9780071635233
Another insight into Toyota, with many useful guidelines. Toyota's problems in recent years do not lessen the validity of *The Toyota Way*: the problems have emerged from not being consistently rigorous with it.

The Universe Story, Swimme, B. and Berry, T. (1992) Harper Collins. 9780062508355
Including an exploration of the Natural Systems principles. Much more to explore at http://www.thomasberry.org

Goals Gone Wild: The Systematic Side Effects of Over-Prescribing Goal Setting, Ordóñez, L. D., Schweitzer, M. E., Galinsky, A. D., and Bazerman, M. H., Harvard Business School, and available from http://hbswk.hbs. edu/item/6114.html as a free download.
A remarkable demolition job on historic practice from one of the foremost institutional advocates.

Variation

Understanding Variation, Wheeler, D. J. (2000, second edition) SPC Press. 9780945320531
This is a short and very accessible book. If you only read one book about variation, make this one it. There are plenty more from Don Wheeler.

Theory of Knowledge

As we have explored in this book, we see the PDSA Cycle as a practical manifestation of the scientific method. There is no shortage of books about science; the most accessible for most readers are ones that tell stories to illustrate the learning process. Any book or TV or radio programme about the Large Hadron Collider at CERN will have multiple references to the power of theories, experiments, predictions and falsifications.

Curiosity: How Science Became Interested In Everything, Ball, P. (2012), Vintage. 9780099554271
A remarkable history book. Curiously, he doesn't really cover science in leadership, but the lessons are there for us nonetheless.

The Faber Book of Science, Carey, J. (Ed.) (1995), Faber and Faber 9780571179015
A collection of stories illustrating the use of science in understanding the natural world.

Contextual Teaching and Learning, Johnson, E. B. (2001), Corwin Press. 978076197865
Elaine Johnson makes a fascinating exploration of the three Natural Systems Principles in the inspiring context of helping children learn.

Psychology

Many of the books we recommend integrate thinking about psychology into their message, but Alfie Kohn addresses the relevant aspects of the subject directly.

No Contest, Kohn, A. (1993, second edition), Houghton Mifflin. 9780395631256
An entertaining look at many aspects of human motivation and how much of the practice in employment and families is entirely counter-productive. If you like this you will also like his book *Punished by Rewards.*

A Beginner's Guide to the Brain, Johnson, E. B. (2012), The Teaching and Learning Compact. 9780983096513
A brief and readable inspiring synthesis of the vast amount of new knowledge about the brain gained over the last few years.

Human needs and motivation: Abraham Maslow and Frederick Herzberg reported over fifty years ago that money was not the motivation for people's satisfaction in life once their basic needs were attended to. Much more about their original work, and studies that validate it over the decades, is available on the web under their names.

Guides for Improvement and Management Practitioners

Peter Scholtes wrote two complementary classics: *The Team Handbook,* (with others; 2003, third edition) Oriel. 9781884731266, and *The Leader's Handbook*, (1998), McGraw Hill. 9780070580282

The Improvement Guide, Langley, G. J., Moen, R., Nolan, K. M., Nolan, T. W. Norman, C. L and Provost, L. P. (1996), Jossey Bass. 9780787902575 A formal, detailed reference book, based upon Deming's approach.

The Process Manager's Handbook, Gillett, J. and colleagues (2012, fourth edition), Process Management International 9780954060527 PMI's summary of the principal methods and tools, and *The Project Leader's Handbook,* (2011), Process Management International 9780954060541 – our companion to *The Process Manager,* expanded to include many improvement project tools for teams and individuals.

Awards bodies

Deming Prize: www.juse.or.jp/e/deming
Malcolm Baldrige National Quality Award: www.baldrige.org
EFQM Excellence Award: www.efqm.org

Index